Microalgae

Current Research and Applications

Edited by:

Maria-Nefeli Tsaloglou
Department of Chemistry and Chemical Biology
Harvard University
Cambridge, MA 02138, U.S.A.
and
Ocean and Earth Science and Institute for Life Sciences
University of Southampton
Hampshire, U.K.

Caister Academic Press

www.caister.com

ISBN: 978-1-910190-27-2 (paperback)
ISBN: 978-1-910190-28-9 (ebook)

Contents

Current books of interest

- Climate Change and Microbial Ecology 2016
- Biofilms in Bioremediation 2016
- Gas Plasma Sterilization in Microbiology 2016
- Virus Evolution 2016
- Arboviruses: Molecular Biology, Evolution and Control 2016
- *Shigella*: Molecular and Cellular Biology 2016
- Aquatic Biofilms: Ecology, Water Quality and Wastewater Treatment 2016
- Alphaviruses: Current Biology 2016
- Thermophilic Microorganisms 2015
- Flow Cytometry in Microbiology: Technology and Applications 2015
- Probiotics and Prebiotics: Current Research and Future Trends 2015
- Epigenetics: Current Research and Emerging Trends 2015
- *Corynebacterium glutamicum*: From Systems Biology to Biotechnological Applications 2015
- Advanced Vaccine Research Methods for the Decade of Vaccines 2015
- Antifungals: From Genomics to Resistance and the Development of Novel Agents 2015
- Bacteria-Plant Interactions: Advanced Research and Future Trends 2015
- *Aeromonas* *2015*
- Antibiotics: Current Innovations and Future Trends 2015
- *Leishmania*: Current Biology and Control 2015
- *Acanthamoeba*: Biology and Pathogenesis (2nd edition) 2015
- Microarrays: Current Technology, Innovations and Applications 2014
- Metagenomics of the Microbial Nitrogen Cycle: Theory, Methods and Applications 2014
- Pathogenic *Neisseria*: Genomics, Molecular Biology and Disease Intervention 2014
- Proteomics: Targeted Technology, Innovations and Applications 2014
- Biofuels: From Microbes to Molecules 2014
- Human Pathogenic Fungi: Molecular Biology and Pathogenic Mechanisms 2014
- Applied RNAi: From Fundamental Research to Therapeutic Applications 2014
- Halophiles: Genetics and Genomes 2014
- Molecular Diagnostics: Current Research and Applications 2014
- Phage Therapy: Current Research and Applications 2014
- Bioinformatics and Data Analysis in Microbiology 2014
- The Cell Biology of Cyanobacteria 2014
- Pathogenic *Escherichia coli*: Molecular and Cellular Microbiology 2014
- *Campylobacter* Ecology and Evolution 2014
- *Burkholderia*: From Genomes to Function 2014
- Myxobacteria: Genomics, Cellular and Molecular Biology 2014
- Next-generation Sequencing: Current Technologies and Applications 2014

www.caister.com

Preface

This book is a dedicated volume on the most recent advances in phytoplankton research. The aim of the book is to introduce the emerging and exciting field of phytoplankton research, and to provide information on the newest developments and applications by leading scientists from around the world.

It will be a useful manual to postgraduate students, as well as to more senior researchers. Although some background knowledge is required, an authoritative chapter on methods of cultivation is included. This provides practical advice, which could prove useful to even the most novice of scientists.

The editor would like to, first and foremost, thank the authors of the book, who gladly volunteered their expertise and made an idea into reality. Finally, the editor would like to acknowledge the patience and support of her partner during long nights and weekends.

Maria-Nefeli Tsaloglou

1

Introduction

Maria-Nefeli Tsaloglou[1,2*] and Carmelo R. Tomas[3*]

[1]Ocean and Earth Science and Institute for Life Sciences, University of Southampton, Southampton, Hampshire, SO17 1BJ, U.K. M.Tsaloglou@soton.ac.uk
[2]Chemistry and Chemical Biology, 12 Oxford St., Harvard University, Cambridge MA02138, US. Tsaloglou@gmwgroup.harvard.edu
[3]Center for Marine Science, University of North Carolina Wilmington, 5600 Marvin K. Moss Lane, NC 28409-5928, U.S.A. tomasc@uncw.edu
*Both authors for correspondence.

An understanding of the species composition and abundance of phytoplankton is essential for defining aquatic ecosystems, and their potential changes. Morphotaxonomy has been useful in the past two centuries in describing the phytoplankton communities. A key element of this method is the ability to grow phytoplankton, which are first collected from the field, in a lab environment. Chapter 2 reviews methods in cultivation of phytoplankton. The authors also discuss necessary growth media and division rates for different algal species.

However, morphotaxonomy shows certain limitations, mainly resulting from the inherent variability of cells, as well as the unreliability of the morphological features between algal species. Since arduous and long training periods are needed to acquire the skillset required for morphotaxonomy, entire phytoplankton compartments have essentially been ignored or not considered. The reason for this disregard has largely been the lack of trained morphotaxonomists. However, methodological advancements in observations have aided the advancement of the field. These include optical techniques (bright-field and confocal microscopy and their accessory techniques, electron microscopy, both SEM and TEM), and automated instrumentation, such as flow cytometry. Flow cytometers can detect and quantify size fractions, which were previously too small to visualize by microscopy. And so, the "*promise of morphology*" may have reached its limits but remains as an essential means of ground truthing species.

The drawbacks of morphotaxonomy have led to the next advancement in microalgal research. This is the "*molecular promise*"; with the alluring feature that genetics could provide us with unequivocal information on speciation. Molecular skills are relatively easier to acquire compared to morphotaxonomy. Molecular methods are remarkably successful in defining the larger picture of phylogeny at the kingdom and division level, but are often not quite as effective at the species level. The area of molecular methods for phytoplankton is still in its early stages, compared to morphotaxonomy, and we are only now beginning to discover their limitations. For instance, we are still discovering which are the definitive genes to sequence and analyze. In many dinoflagellates, no clear single target sequence emerges for identification of the entire taxon but the fallback has to be on consensus sequences, or non-nuclear-encoded genes (chloroplasts, mitochondria, *etc.*). To some extent, an extremely sensitive molecular method is used in a shotgun fashion to define taxa. This disadvantage has more specific implications to species-specific probes (*e.g.* molecular beacons, FISH probes). Such probes are accurate for identification of microalgae in large natural populations, but could be used interchangeably for the better-known species. For example, molecular probes best identify the dinoflagellate genus *Gambierdiscus*. Conversely, emerging "complexes" such as *Raphidophyte* genus *Chattonella* are yet to be distinguished by common genes, whereas they are distinctly different species by morphology. Do we need more genes to be tested? It may equally erroneous to say that if the genes can't differentiate them, then perhaps the distinct morphological differences are not significant? The same argument may be applied to the dinoflagellate genus *Alexandrium*. Methods and approaches to study gene expression in phytoplankton, in general, are reviewed in Chapter 3. The chapter also provides with a short overview of how these molecular methods are used to study the effect of nutrients to gene regulation of phytoplankton in the oceans.

The dichotomy between molecular and morphological identification is critical with the "*harmful algal paradigm*". In this case, accurate species confirmation not only has an esoteric academic aspect to it, but also implies a deeper meaning involving species that can alter ecosystems. The presence of a toxic alga can reflect a threat to public health, environmental stability and ecosystem dystrophy and, therefore, defining a harmful species accurately is critical for matters beyond the immediate area of taxonomic studies. Normalization of biomass to chlorophyll, or use of undefined taxonomic units as was needed in some mathematical simulations is not sufficient. Harmful algal blooms and their effects to local environments, especially from a European perspective, are reviewed in Chapter 4.

Chapters 5 and 6 review two exemplary categories of microalgae, *Emiliania huxleyi* (EHUX) and bioluminescent species, respectively. EHUX is a globally abundant species that contributes ~50% of the carbonate flux on the ocean's surface by way of sinking calcite cocoliths, and is thus critical in ocean biogeochemistry. Bioluminescent microalgae can produce blooms, in some cases harmful. Most importantly, they produce bioluminescence using physiological mechanisms, which remain largely unknown. Finally, an alternative for phytoplankton identification in the field, which requires neither morphotaxonomical nor molecular skills, is the use of *in situ* sensors. Chapter 7 provides an overview of the state-of-the-art in microfluidic and *in situ* sensors for phytoplankton analyses. These sensors can increase the number of sites, as well as the frequency of monitoring in the oceans, which are largely under-sampled.

It is clear that we need the best of all possible worlds. We still need skilled morphotaxonomists who can use all the tools available, and apply each as needed for the problem species at hand. At the same time, the skills of molecular taxonomists are becoming increasingly indispensable. Molecular experts are necessary to evaluate the use of defining genes, and to excel in techniques for analyzing the sequences to illustrate phylogenetic relationships. As a new addition to the field of phytoplankton research, skilled engineers, who can integrate lab-based methods and develop field-use solutions, are now part of the community. The field is now largely multi-disciplinary, and this comes with its own challenges.

Molecular skills are relatively easy to acquire. A competency in a given molecular method can be acquired with time spent in critical workshops, and some time invested in working with known species. On the other hand, the morphotaxonomic approach does require long hours at microscopes, carefully looking at fine details, and then becoming skilled with the first line of observations, the bright-field microscope. It takes extensive training to become adept in the accurate identification of the myriad of species encountered in natural samples, while the world mentors, of these skills are fast disappearing. Training is becoming more and more difficult to acquire and, in some areas of the world, students often have to travel abroad to a international workshops and special training courses. For the more applied aspects of the phytoplankton research, which require engineering skills, attracting skilled mechanical and electronics engineers is key. This can be particularly challenging as they could have alternative careers with significantly higher enumeration than research for oceanographic applications.

To conclude, national funding agencies need to address these challenges set by the skillset required for phytoplankton research. Newer approaches have to be supported, without abandoning the more traditional skills, like morphotaxonomy. Deliberate and difficult choices will have to be made in order to promote research for phytoplankton at this critical time for emerging environmental health issues, impacts of climate change and the better need for balance in the questions of algal species.

Other Publications of Interest

Flores, E. and Herrero, A. (2014). The Cell Biology of Cyanobacteria. Caister Academic Press, U.K. ISBN: 978-1-908230-38-6

Herold, K.E. and Rasooly, A. (2009). Lab-on-a-Chip Technology: Fabrication and Microfluidics. Caister Academic Press, U.K. ISBN: 978-1-904455-46-2

Herold, K.E. and Rasooly, A. (2009) Lab-on-a-Chip Technology: Biomolecular Separation and Analysis. Caister Academic Press, U.K. ISBN: 978-1-904455-47-9

Wilkinson, M.G. (2015). Flow Cytometry in Microbiology: Technology and Applications. Caister Academic Press, U.K. ISBN: 978-1-910190-11-1

Xuefeng, L. (2014). Biofuels: From Microbes to Molecules. Caister Academic Press, U.K. ISBN: 978-1-908230-45-4

2

The Cultivation of Marine Phytoplankton

Maria G.S. Jutson[1], Richard K. Pipe[1*] and Carmelo R. Tomas[2]

[1]The Marine Biological Association of the United Kingdom, The Laboratory
Citadel Hill, Plymouth, Devon, PL1 2PB, U.K. richardkpipe@gmail.com
[2]Center for Marine Science, University of North Carolina Wilmington, 5600 Marvin K. Moss Lane, NC 28409-5928, U.S.A. tomasc@uncw.edu
*Author for correspondence.

Contents

Abstract
The isolation and cultivation of phytoplankton can be an essential, and also satisfying, process for studying microalgae, in detail. Most species reproduce asexually, therefore establishment of unique strains in culture is relatively straightforward. Within the context of this chapter, the term strain will be used to define a genetically homogenous clone propagated from an individual microalgal cell. In essence, individual cells are isolated and placed in a suitable environment for growth. However, the actual process of establishing permanent cultures involves many steps and can entail prolonged periods of time. The validity of using cultured cells, as a model for physiological function, requires that culture conditions should resemble the natural environment as closely as possible. This chapter will review methods for cell isolation, sterilization and culture maintenance conditions. We will go through the different types of culture methods, as well as the many and varied kinds of cell enumeration, which is key to culture maintenance.

Introduction
Historically, algal cultures have been grown on a small scale in research laboratories and universities for fundamental experiments and feeding other marine organisms. Most commercial growers have cultured a limited range of species principally as dietary supplements, for production of pigments, or as food for aquaculture. Commercial production of algae is a small volume, high value market (Hennenberg et al., 2009) involving *Chlorella, Spirulina, Dunaliella, Haematococcus* and additional species for aquaculture (Meng et al., 2009).

Prior to delving into the details of cultivation, a clear idea regarding the reasons for culturing microalgae needs to be established. To answer taxonomic questions, laboratory cultures may be essential for detailed morphological assessments. Other physiological questions on life cycle stages, including 'cysts', may require cultivated species. Detailed analyses of bioactive compounds, or pigments, also necessitate the production of a large biomass of the microalgal species from which these compounds can be isolated.

Microalgal cultures have many potential uses, including:

- Cell morphology and life cycle studies
- Physiological experiments

- Molecular genetics
- Aquaculture
- Studies of natural products and toxins
- Synthesis of organic compounds
- Production of nutraceutical and pharmaceuticals
- Carbon dioxide sequestration
- Biofuel production

Isolation of culture cells

Collection of samples

For coastal and oceanic sites plankton is usually concentrated by gentle towing with a plankton-net, e.g. 10 minutes with a 200 µm and a 600 µm mesh net. Specialized water sampling bottles may be used for taking samples of known volume from a precise depth. Most water-sampling bottles consist of a cylindrical tube with stoppers at each end, and a closing device activated by a messenger; they include Kemmerer, Van Dorn, Niskin Bottles and Fjarlie bottles. Some microalgal species may also be collected from rock pools, or the shoreline.

Having collected a phytoplankton sample, gentle handling and rapid transfer times are important factors, as damaged and dying cells will not produce viable cultures. It is essential to make a record of relevant collection details. These could include: location, date, depth of sample, water temperature etc. When transporting collected samples, small plastic drinking water bottles are ideal. The bottles have to be rinsed with the sample water itself and filled so that no bubbles are present. This is particularly important for overnight shipping. The bottle may then be wrapped in wet toweling to insulate the sample. Ideally, a courier should be used as many can guarantee next day delivery. If the sample has been collected using a tow net or is from a very dense bloom, it has to be diluted prior to shipping. To dilute the sample, glass-microfiber-filtered (GF/F) water of the same salinity must be used. Shipping should be avoided during hot periods if possible, or cool packs, separated from the sample with paper or towels, will have to be used. In general, unless the samples are of Arctic or Antarctic origin, ice should not be used during shipment.

Single-cell isolation

This is carried out using a micropipette and is probably the most common method for isolating individual phytoplankton cells. Traditionally a fine glass Pasteur pipette or a glass capillary tube has been used. A finely controlled picking out device, linked to a fine glass pipette, has been developed by Carmelo Tomas, specifically for isolating individual cells. Alternatively a sterile extended fine tip plastic transfer pipette can be used. The aim of micropipette isolation is to pick up a single cell from the sample and deposit it, without damage, into a sterile droplet of medium. The sterile droplet containing the target cell, and probably other cells, is examined. This can be done using a multi-test slide. Another useful method involves using a tissue-culture inverted microscope with the aid of welled slides or welled plates. Slides can be purchased with cover glass-bottoms for critical observations and photographs, and these can be used with 40x and even 100x objective lenses. When using any of these slides, pulled glass pipettes bent at 90° are useful. Standard upright microscopes may not be very comfortable because of the small working distance of the objective lens and sample. Good quality dissecting microscopes are commonly used and can be very effective if the sample being selected is a larger one (>15 µm). The down side of this type of microscope is the limited magnification (usually < 200x total) and good lighting may be difficult to obtain. Washing and transferring cells from well to well in a microtiter plate or multi-test slide can be easily done. This process needs to be carried out enough times to ensure clean isolation of a single cell but without causing cell damage by excessive handling. Cells in specific plates can be observed day after day without disturbing them, by simply placing the plate under the inverted microscope and observing them in a closed plate. Plates can be sealed with tape to reduce evaporation.

A newly isolated strain can be difficult to maintain so the optimum culture medium in which the alga will grow needs to be established. A reasonably reliable way of doing this is to simply put the isolates into several culture media and monitor growth. Knowledge about the growth medium used for closely related algae is often valuable. Daily microscopic examination should be carried out. Once the culture is growing it can be transferred to a 100 mL glass flask or plastic tissue culture vessel. When the culture is established, it can be grown in larger volumes.

Other methods of isolation include: the use of agar, dilution techniques, centrifugation, flow cytometry and phototaxis (Andersen, 2005).

Sterilization

Filtration

Filtration is mostly suitable for small volumes and heat labile components such as vitamins. Membrane filters with pore size of 0.2 µm should remove contaminating organisms including bacteria; however viruses and fungal spores may still pass through. A 0.1-µm filter can be used although filtration of large volumes will take a long time. A wide range of filter units and membranes are available including both autoclavable and single-use systems. Whilst filter sterilization can be convenient it should be used with caution as filters used may leak compounds into the filtrate and absolute sterilization is not guaranteed (Stokner et al., 1990).

Autoclaving

Autoclaving involves using high-pressure saturated steam at 121°C to sterilize equipment and media. It is the standard microbiological system used as it guarantees sterility, including eradication of viruses and spores. Both large and small commercial autoclaves are generally used although for small volumes pressure cookers may also be suitable.

When autoclaving liquids, it is essential that containers are only ¾ full, with screw caps partially loosened to prevent excess build-up of pressure. Under autoclave conditions the pH of seawater is raised as CO_2 is driven out of solution, which may cause precipitation of salts. Precipitates should generally to be avoided as they tie up nutrients, metals and organics in a particulate form thus changing the availability of these components of the media. However, arguments have been put forward to suggest that precipitates are not problematic as they slowly release nutrients into the media over time and thus serve as a nutrient reservoir. To minimize the formation of precipitates in seawater or media, autoclaving may be carried out in Teflon bottles therefore avoiding adhesion of ions to the walls of the bottle. Precipitation may also be avoided by the addition of distilled water, usually ~5%, however salinity may be affected.

When autoclaving glassware and plasticware the high temperatures and pressure used will cause the release of substances from the vessel walls. Borosilicate glass such as Pyrex® may release metals that have been used during manufacture. Polycarbonate bottles may continue to release plastic stabilizer for several months during the autoclave process. Some of these substances, known as phthalate esters, are toxic and can interfere with the growth of sensitive species. When autoclaving plastic-ware it is a good idea to have a small amount of deionized water in the container that will capture most of the dissolved plasticizers (phthalates). This can then be discarded under aseptic conditions just prior to filling the container with sterile medium. Autoclave tape is useful to indicate that the items have been treated and are sterile.

Dry heat

Dry heat is generally carried out in a hot air oven and is useful for sterilizing glassware (tubes, flasks, beakers etc.) and glass pipettes. Glassware to be processed in this manner needs to be cleaned, dried and openings covered with aluminium foil to preserve sterility on removal from the oven. Items should be placed in an oven at ~160°C for a minimum of two hours. Pipettes stuffed with non-absorbent cotton can be placed in glass canisters or stainless dishes with lids although care must be taken not to heat for too long, otherwise the cotton may burn.

Pasteurization

Pasteurization is a gentle and often effective method of sterilizing seawater and media; it can be a time-consuming process. It is particularly useful for sterilizing liquids which should not be autoclaved or exposed to temperatures >100°C. In general, it involves heating a liquid to a specific temperature for a predefined length of time and then immediately cooling it after it is removed from the heat. For seawater and media, it has been shown that heating to 80-90°C and maintaining at that temperature for an hour is effective. This process needs to be repeated a second time before the seawater or media are ready for use. Pasteurization may be accomplished in a temperature regulated water bath with either a thermometer or thermistor to assess the temperature within the liquid. The advantages of pasteurization are that it is gentle, normally destroys those organisms commonly found at lower temperatures and organic compounds remain intact. Delicate phytoplankton cultures normally respond favorably to pasteurized media. The disadvantages are that the processing time is normally longer than autoclave sterilization and some organisms (viruses and some bacteria) may not be completely destroyed.

Ultraviolet treatment

Ultraviolet treatment is not generally used for the preparation of culture media as it is not considered a sufficiently reliable technique for sterilization. UV light does not penetrate water uniformly and microbes may attach to vessel surfaces potentially rendering sterilization ineffective. For ultraviolet treatment to be effective, the UV light should be of sufficient strength as bulb strength varies with time. The water should be kept in quartz glass tubing, which is transparent to UV light and maintained in the UV light stream for the required time to destroy all microbes.

Microwaving

Microwaving has the advantage of speed for sterilizing small volumes (1-1½ liters) where it has been shown to kill bacteria and fungi within 10 minutes (Keller et al.1988). However, as the power, and efficiency, of microwave ovens varies a great deal each unit will require protocols to be developed. In general, this method is not practical for large volumes but is useful for laboratories that do not have autoclaves.

Culture maintenance conditions

Temperature

Ideally, temperature should reflect the conditions from which the algae were isolated. Therefore, polar isolates should be maintained at <10°C, temperate isolates at 10-25°C and tropical isolates at >20°C. Temperatures above 30°C are usually fatal for marine algae. Temperature stability should be kept to ±2°C, where possible. Incubation temperatures higher than 20°C may be combined with increased light intensities to prevent photo-inhibition or damage.

Lighting

Traditionally, cultures are housed in a constant temperature room with a large north-facing window (in the northern hemisphere), usually with supplementary lighting. However, very few facilities have the luxury of natural lighting and most have to rely on indoor lighting provided by various lighting systems. One most commonly used system employs cool white fluorescent bulbs to give the adequate flux density to grow cells properly. Various bulbs are available, some with spectra designed for terrestrial plants; however, these are expensive and give no advantage over the plain cool white bulbs. A recent innovation is the advent of light emitting diodes (LED) that now come in a variety of power strengths and spectra, similar to those of other sources. These bulbs are presently more expensive than the fluorescent ones but consume a fraction of the electricity and normally have a 50,000-hour burn resulting in a significant savings over more conventional bulbs. When using fluorescent tubes they should be placed at a distance from the cultures, so as not to cause any overheating of the algae. Photoperiods are required for the maintenance of most algae. Continuous light can be detrimental and may kill some algae. In most culture collections the light/dark regimes vary between 12:12 and 16:8 hour light/dark. The regime used will depend on whether the cultures are kept in a room where natural light is available or not and, if so, the time of year. Direct sunlight should always be avoided. Over-illumination is a widespread mistake in the maintenance of cultures. Problems include localized heating of algae, photo-oxidative stress and bleaching (loss of pigmentation) effects.

Culture media

Media for the culture of marine phytoplankton in the laboratory can be either defined or undefined. Defined media use high quality deionized water to which salts are added, together with the nutrients required to make the formulation of choice. Physiologists (who control all variables as closely as possible) generally prefer this type of media. The general finding is that defined media will not grow all species and sometimes result in species growing abnormally. Therefore, defined media are generally best used for specific needs but not for maintaining cultures for long periods of time.

Undefined media consist of a seawater base, either natural or artificial, which is then supplemented with additional nutrients. The media are best prepared using good quality, oligotrophic seawater collected some distance from land to avoid runoff contamination. Collection of seawater needs to be done carefully with pumps and tubes that are non-metallic so as to avoid contamination. The quality of the seawater may be improved by ageing for a few months to allow for bacterial degradation of inhibitory substances. It is usually kept in seawater-conditioned polyethylene carboys or tanks and stored in a cool, dark place to discourage growth of photosynthetic organisms. Proprietary brands of artificial seawater may be used if natural seawater is not readily available. Seawater must be filtered prior to use to remove particulate matter. Pre-filtering (e.g. Whatman chromatography filter paper no. 1) can be performed to remove larger particles before further filtration through 0.45-μm filter.

The formulation of each medium will have its own specific nutrient composition. The macronutrients, nitrogen and phosphorous, are major additional components of all culture media recipes. Nitrate is the preferred form of additional nitrogen, although a few recipes use ammonium. Organic nutrients are another consideration and for nitrogen, a variety of amino acids have been used. Inorganic orthophosphate is the usual form of additional phosphorous, although organic glycerophosphate can be used. Some algal groups including diatoms, silicoflagellates and some chrysophytes also have a requirement for silica in the culture medium. Recipes vary but these macronutrients are essentially the same for different media for example the F series (f/2, K, L1 etc). The components, which do vary in the different media, are the trace metals and perhaps particular vitamins. Low concentrations of some trace metals are essential for algal growth. These trace metals include; iron, zinc, copper, cobalt, manganese and nickel. Vitamins are routinely added to nearly all formulations and consist of a combination of thiamin (B_1), biotin (H) and cobalamin or cyanocobalamin (B_{12}). Concentrations of trace metals and vitamins vary between different media recipes but are usually in the range 10^{-7}–10^{-12} moles. These components will usually need to be made up in two stages. Primary stocks are made up which are then diluted into a working or secondary stock. These primary stocks are long lived and can be kept frozen at -20°C for extended periods of time.

There are several nutrients of particular note worth mentioning that have special usages. One of these is the micronutrient selenium. Both K and L1 medium have this as an important component. Although the exact physiological function of selenium is not well understood, it was not until the inclusion of this metal in media that hard to grow benthic or open ocean species were successfully cultured. The form of selenium is important and should be the reduced form. The common reagent used is selenous acid (H_2SeO_3) and is used at 10^{-8} molar concentration. Iron is another critical element that offers some difficulty in preparation. For many years, the inorganic form of iron as ferric chloride ($FeCl_3.6H_2O$) was used but this may precipitate within a short time, leaving the amount of iron in solution unknown. Even in a precipitated solution, the amount of iron added in the specified volume is generally considered to be far beyond the requirement. However, an alternative to ferric chloride is a chelated form of iron as sequestrene (di-ferric sodium EDTA salt). This form of iron is easily soluble, remains in solution longer than ferric chloride and is commonly used as a higher plant supplement. Historically, the use of soil enrichment extract was found to be essential for the growth of many algal cultures (Pringsheim, 1946). Soil extract enriched medium is still used successfully by some culture collections today. The most suitable type of soil to use is sieved sandy loam, which is lime-free and has not been treated with any fertilizers and pesticides.

The preparation of different culture media varies considerably, however the final solution needs to be sterile. Sterilization can be of each component separately or alternatively the complete medium once prepared. There are many methods utilized to accomplish sterile culture media including: filtration, autoclaving, pasteurization, and ultraviolet treatment.

pH

The control of pH in culture media is important, as certain algae will only grow within narrowly defined pH ranges. The pH of natural seawater is usually around eight. Because of the natural buffering capacity of seawater it is quite easy to maintain the pH of marine culture media. The buffer system may be overwhelmed during autoclaving, when high temperatures drive CO_2 out of solution and hence cause a shift in the bicarbonate buffer system and thus an increase in pH. As culture medium cools after autoclaving, atmospheric CO_2 enters the solution but if normal pH is not fully restored the level will need to be adjusted. Alternatively, the pH of seawater may be lowered prior to autoclaving (adjustment to pH 7-7.5 with 1M HCl) to compensate for subsequent increases.

Some media recipes include the addition of extra buffer, either as bicarbonate, tris-hydroxymethyl-aminomethane (Tris), or glycylglycine to supplement the natural buffering system. Glycylglycine is rapidly metabolised by bacteria and hence should only be used with axenic cultures. These additions are generally not necessary if media are filter-sterilized, unless very high cell densities are expected.

Very dense cultures of microalgae may take up sufficient CO_2 to also raise pH level. The problem of CO_2 depletion in dense cultures may be reduced by having a large surface area of media exposed to the atmosphere relative to the volume of the culture, or by bubbling with either air (CO_2 concentration ~0.03%) or air with increased CO_2 concentrations (0.5 to 5%). Unless there is a large amount of biomass taking up the CO_2, the higher concentrations could actually cause a significant decline in pH levels. When bubbling is employed, the gas must first pass through an in-line 0.2-µm filter unit to maintain sterile conditions.

Aeration

For many microalgal species, aeration is not an option since the physical disturbance may inhibit growth or kill cells. However, it may be necessary when cells must be kept in suspension in order to grow. This is particularly important for heterotrophic dinoflagellates. In concentrated cultures, aeration may also be necessary to prevent nutrient limitation effects due to the stacking of cells, and to increase gas diffusion. It should be noted that in the ocean cells seldom experience turbulence, hence mixing should be gentle. Aeration methods include: bubbling with air, gentle manual swirling, using a plankton wheel or roller table. Most cultures do well without aeration, particularly when not too concentrated, but when possible gentle manual swirling, once a day, is recommended.

Culture vessels

Containers for culturing algae should be non-toxic, enable illumination and be capable of withstanding sterilization. Borosilicate glass has been used traditionally, but is increasingly being replaced by plastic-ware, as predicted by Pringsheim back in 1946. Glass vessels are relatively inexpensive but are labor intensive to maintain, as they require cleaning and sterilizing. Pre-sterilized, disposable, plasticware can be purchased in ready-to-use packages but is relatively expensive, both economically and environmentally. Plastics can be coated with substances which may enhance growth of many algae. A range of plastics may be suitable for culturing including: Teflon®, polycarbonate, polystyrene, polyethylene terephthalate (PET) and polyethylene terephthalate copolyester glycol modified (PETG).

Plastic vessels are available in a wide range of formats from 96 well microtiter plates to large volume carboys. Most tissue culture flasks are flat, enabling cultures to be grown on shelves. The caps may be solid or can have a porous insert to allow gaseous exchange whilst maintaining sterility. The flasks with porous inserts often give better results for some species. Most sterile plasticware is sterilized by gamma radiation, which does not inhibit growth in the majority of species. Some flasks used in medical procedures are gas-sterilized. These should be rinsed with sterile deionized water prior to use in order to remove any traces of the sterilizing gas.

Erlenmeyer flasks have traditionally been used for maintaining laboratory cultures. Whichever container is chosen it should enable gaseous exchange but prevent contamination and limit evaporation. Various closures may be utilized including; cotton wool, metal or glass lids, screw caps, foam bungs and BugStoppers. Cotton wool plugs should be covered with an aluminium cap or greaseproof paper fixed with a rubber band to reduce evaporation.

Culture methods

Batch culture

An algal inoculum is placed in a fixed amount of complete culture medium and incubated in a favorable environment for growth. Resources are finite when growing algal cultures in a limited volume. When the resources present in the culture medium are abundant, growth occurs according to a sigmoid curve, but once the resources have been utilized by the cells, the cultures die unless supplied with new medium. In practice, this is done by routine sub-culturing, i.e. aseptically transferring a small volume of existing culture in late log/stationary phase to a large volume of fresh culture medium at regular intervals. The characteristic pattern of growth involves; *lag phase* where cells are adapting to their environment and do not divide; *exponential (log) phase* where cells multiply rapidly; *stationary phase* where the number cell division is limited by the reduced nutrients and as a result the rate of cell growth matches the rate of cell death and *death phase* where cells run out of nutrients completely and die.

When sub-culturing aseptic technique should always be adopted and where possible a laminar flow hood used to reduce any possibility of contamination. For routine maintenance, the goal is to sub-culture the alga at the end of its exponential growth phase. This depends on the sensitivity of the strain, with sensitive strains requiring a transfer cycle of 1-2 weeks and the most robust strains only requiring transfer every 2-3 months. At the Plymouth Collection, the majority of strains are sub-cultured every four weeks. The volume of inoculum is dependent upon cell density but is usually in the range of 1-5% of the fresh culture medium. Algae kept on agar slopes can be transferred once every six months.

Continuous culture
Resources are potentially infinite in continuous cultures. Therefore, the algae are maintained at a chosen point on the growth curve by the regulated addition of fresh culture medium. In practice, a volume of fresh culture medium is added automatically at a rate proportional to the growth rate of the alga, while an equal volume of culture is removed. By doing this, the growth limiting substrate is replenished and any growth inhibitory products are removed.

Semi-continuous culture
Cells in an actively dividing state are maintained in culture by periodically draining off a proportion of the depleted medium and replenishing it with fresh medium. This system is not sustainable indefinitely as the build-up of inhibitory or toxic by-products will ultimately result in cell death.

Culture of large volumes
Scaling up of algal cultivation has been carried out in both open-air systems or in specially-developed photo-bioreactors. Commercial cultivation of microalgae is limited to a small number of species and is currently carried out, almost exclusively, in open systems. This is because large ponds and raceways are relatively cheap to construct and operate compared with photo-bioreactors. The majority of commercial open-air systems are sited in tropical or sub-tropical regions where there is less fluctuation in light and temperature. The main disadvantages of these systems are: the risk of contamination by predators or other algae, evaporation, the footprint required and relatively low cell densities obtained. A very large number of closed photo-bioreactor systems have been developed. The main benefits of a closed system are: full environmental control, reduced risk of contamination and a much smaller footprint than open systems. The major disadvantages are expense of construction and high running costs.

Harvesting cultures
In order to remove large quantities of water and process large algal biomass volumes several steps are required including: separating algae from the medium, drying the cells and processing to obtain the desired product. Harvesting methods depend primarily on the type of algae being grown. Common harvesting processes involve flocculation, centrifugation and filtration. Probably the most popular method of harvesting large quantities of culture uses continuous flow centrifugation. These units are made to run continuously and result in a moist paste that can be dried further or stored in cool or

frozen conditions. A 60-liter bag culture can be harvested in less than two hours. All methods should be energy-efficient and relatively inexpensive so selecting easy-to harvest algae is important. Methods for drying and further processing depend upon the product required.

Contaminating organisms

Microalgal cultures are susceptible to a wide range of contaminating microorganisms including: viruses, bacteria, fungi and protozoa. Cultures may be grown either axenically i.e. not contaminated by or associated with other living organisms or xenically i.e. contaminated by or associated with other living organisms. Within the context of algal culturing these terms usually refer to bacteria. The interactions between bacteria and phytoplankton species are complex (Doucette, et al., 1998). When growing microalgae in culture it has been shown that some individual strains of bacteria may inhibit growth whilst others act as a stimulant (Biddanda et al., 2006). However, it is generally agreed that the performance of many algal cultures is better with some bacterial contamination as some microalgal cultures may die when made axenic. This is probably due to the termination of obligate symbiotic relationships with bacteria.

If a biological contaminant appears in a culture, a clean clonal culture may be re-established by isolating a single cell from the contaminated culture, using a micropipette. Alternatively, the culture can be streaked on an agar plate in the hope of attaining a colony free of contaminants. Neither of these methods work particularly well for eliminating bacteria attached to the surface of the algal cells. However, placing a test tube of microalgal culture in a low-intensity 90 kilocycles/sec ultrasonic water bath for varying lengths of time (a few seconds to tens of minutes) can sometimes physically separate bacteria without killing the algae, making it easier to obtain an axenic culture by micropipette isolation.

In order to achieve an axenic culture it is often necessary to use antibiotics. One of the most successful methods involves using an actively growing culture of algae and exposing it to a mixture of penicillin, streptomycin and gentamycin for around 24 hours. This drastically reduces the growth of bacteria whilst allowing the microalgae to continue to grow. The chances of obtaining an axenic culture when using micropipette or agar streaking isolation methods are then greatly increased. Different algal species tolerate different concentrations of antibiotics, so a range of concentrations should be used (generally 50-500 mg/l). Other antibiotics that can be used include: chloramphenicol, tetracycline and bacitracin. Antibiotic solutions should be made with distilled water and filter-sterilized (0.2-µm filter units) into sterile tubes and may be stored frozen until use. Another approach is to add a range of antibiotic concentrations to a number of subcultures and then select the culture that has surviving algal cells but no surviving bacteria or other contaminants.

Sterility of cultures can be checked by microscopic examination, either with or without a fluorescent dye such as DAPI. Alternatively a small amount of sterile bacterial culture medium (eg. 0.1% peptone) can be added to a microalgal culture and observed regularly for bacterial growth. Absence of bacterial growth does not, however, ensure that the microalgal culture is axenic, since the majority of bacteria do not respond to standard enrichments. In reality, there is no way of demonstrating that a microalgal culture is completely axenic. In practice, therefore, axenic usually means 'without demonstrable unwanted prokaryotes or eukaryotes'.

Fungal contamination is generally worse than bacterial contamination because fungi are harder to eliminate by either physical or chemical methods. If fungal contamination is suspected it can be checked for using peptone or malt extract dissolved in the algal culture medium. Current methods for controlling fungal contamination rely on using aseptic techniques when handling cultures. If cultures become contaminated, physical separation of algal cells from fungal cells, on agar plates may be carried out using a microscope. Some algal strains show positive phototaxis and this ability may also be used to physically clean cultures. A potentially convenient and effective control measure is to add a fungicide to solid media to either prevent fungal contamination or to clean up contaminated cultures. Fungicides which have been used successfully include: carbendazim, thiophanate-methyl, and benomyl (Mahan et al., 2005). These fungicides are considered broad-spectrum fungicides of the benzimidazole family. Cultures may be cleaned by streaking the contaminated algae onto agar plates, containing 40 µg/mL of the fungicide. The algae can be re-streaked onto fresh agar after 5–7 days of growth. This process may need to be repeated several times for heavily contaminated cultures.

Contamination of laboratory cultures with protozoa is not usually a problem when aseptic techniques are employed and cultures can be re-isolated relatively easily. Whereas problems may arise in cultures containing unwanted microalgae such as diatoms, which grow very quickly and can often outcompete other algae. Diatoms can be eliminated with the addition of germanium oxide to the culture (Lewin, 1966).

Cell counting

In order to determine the concentration of cells in a culture it is necessary to count the cells in a given volume. There are several methods available for counting algal cells. These may direct or indirect and range from simple microscope counting chambers to sophisticated electronic devices. The most accurate is to actually see and count cells with a microscope and various chambers that make the counting volume accurate for extrapolation to larger volumes. However, the choice of equipment depends on several factors, including: culture density, size and shape of the cells and presence of extracellular material such as mucilage. While the indirect methods may seem less tedious, in that a surrogate is used for biomass (chlorophyll, fluorescence, optical density, electronic particle counting) the major drawback is that these methods need to be carefully calibrated against standards with cell counts. In addition, by using relative fluorescence or particle density, there is no assurance that only cells are being counted or measured. These counted samples should at least occasionally be verified by direct observation. Variables including, chlorophyll per cell, detrital load and coincidence in particle counting should be of concern when using indirect methods.

Sedgwick-Rafter counting cell

The chamber is rectangular (50x20 mm), 1 mm deep, with an area of 1000 mm^2 and a volume of 1.0 mL. It is best used for relatively large cells, 20-500 µm with cell densities ranging from ~30 per mL up to about 10^5 per mL. The base is divided into 1-mm squares.

A cover glass traps liquid to the correct depth. The chamber must be completely filled with no air bubbles present. Using an inverted microscope or low power compound microscope, cells in each cubic millimeter can be counted. Where cells lie on the lines, only those organisms on two sides of the square are counted. The chamber and cover-slip must be perfectly clean; this is usually achieved by rinsing in ethanol or methanol. Fixation is necessary for motile cells; this is commonly done using Lugol's iodine or alternatively a formaldehyde based fixative. For Lugol's iodine, 10 g potassium iodide, 5 g crystalline iodine are dissolved in 100 ml distilled water. Once dissolved, one drop (~0.05 mL) per 5 mL of algal sample is added to the solution.

Haemocytometer

The haemocytometer has two chambers each with nine 1mm squares with several layers of subdivision. One of the most common rulings is the 'Improved Neubauer', which comprises a grid (3 mm x 3 mm), divided into nine 1-mm^2 areas, with a depth of 0.1 mm. Another useful chamber is the Fuchs-Rosenthal, which also has a total ruled area of 9 mm^2 but has a depth of 0.2 mm, making it useful for bigger cells. However, even with a depth of 0.2 mm, the size of cells, which can be counted accurately, is limited to around 70 µm, as larger cells will seldom form an even distribution in the chamber. Motile cells will need to be fixed when using this method. The chamber and cover-slip must be perfectly clean; this is usually achieved by rinsing in ethanol or methanol. The cover-slip is firmly fixed to the chamber by cohesion as shown by the presence of diffraction rings. It is essential that the cover-slip is symmetrically placed over the double rulings to prevent non-random cell distribution. The cell suspension is transferred to the chamber using a fine pipette. The chamber must be filled using a single application, without overflowing into the channels and must not contain bubbles. The cells must be allowed to settle prior to counting. For cells touching the center line of the triple ruling it is important that only two sides of the square are included in the count and cells touching the center line of the other two sides are excluded. To maintain an acceptable degree of error, at least 50 cells should be counted. To calculate cell density:

Cells per µL= number of cells counted x 1/area x 1/depth x dilution

Other haemocytometer rulings are available including; Thoma, Burker, Malassez, Nageotte and Petroff-Hausser.

Palmer-Maloney Slide

The counting chamber has no rulings, is circular, 17.9 mm in diameter x 0.4 mm deep, with a volume of 0.1 mL. The chamber is provided with two charging channels, 2 mm wide x 5 mm long. To use the Palmer-Maloney counting chamber, a cover-slip is placed on the edge of the chamber and the sample introduced, as the sample fills the chamber, it draws the cover-slip over the chamber. The Palmer Maloney is useful for organisms with a size range of 5–150 µm.

Secchi disc

A Secchi disc is a standard disc, either plain white or with alternating black and white quadrants, which is lowered into the sample until it is no longer visible. This depth of disappearance, called the Secchi depth, is a measure of the turbidity of the water. The Secchi depth is proportional to the algal cell density. Once calibrated using direct cell counting it can be useful for estimating growth of a microalgal culture. The Secchi disk readings can only provide an estimate of turbidity, because errors can arise due to factors like lighting conditions or individual perception. However, it is an inexpensive and straightforward method of measuring water clarity, and thus estimating cell density. Because of the potential for variation between operators methods should be standardized as far as possible.

Spectrophotometer

The use of a spectrophotometer to measure cell concentration provides an indirect method that correlates the algal density with light absorbance at specific wavelengths. Spectrophotometers can use either cuvettes or microplates, which enable cell density to be measured in a wide range of volumes. The method can be reliable and also easy to setup for automatic monitoring systems (Rodrigues et al., 2011). The drawback is the limited accuracy because the cells are not counted directly as the machine measures light, not cells. This, combined with the stochastic nature of liquid cultures, enables only an estimation of cell numbers. The spectrophotometer can also be used to measure the chlorophyll concentration in cultures. Chlorophyll A constitutes approximately 1-2% (dry weight) of algal biomass, which makes it a convenient indicator of algal biomass. However, chlorophyll content in an algal cell is not constant and varies with the nutritional state of the cell. This will obviously affect the accuracy of the cell density estimates derived.

Coulter counter

The machine has one or more micro-channels separating two chambers containing electrolyte solutions. As culture medium containing the algal cells passes through the micro-channel, each algal cell causes a brief change to the electrical resistance of the liquid. The counter detects these changes in electrical resistance and thereby counts each cell. The advantages of using a Coulter Counter are (1) the accuracy it provides and (2) the relative simplicity, enabling large numbers of samples to be counted efficiently. The disadvantages are (1) it does not discriminate between algal cells and other particles, (2) a dense culture needs to be diluted to get an accurate count and (3) it is relatively expensive.

Flow cytometer

The most sophisticated and expensive method for counting cells is to use a flow cytometer. The culture containing the algal cells flows in a narrow stream in front of a laser beam and a light sensor detects the light reflected from the cells as they hit the beam. Flow cytometers can also analyze the shape of cells, their internal and external structures as well as measuring specific components of the cells. The expense and complexity of flow cytometers means that they are rarely used for the sole purpose of counting cells. Flow cytometers incorporating a cell sorter may also be useful for making rapid single cell isolations.

Cryopreservation

The purpose of cryopreservation is to maintain cells at extremely low temperatures whilst maintaining viability. The majority of cryopreservation protocols for microalgae use a two-step controlled rate freezing methodology with the aid of cryoprotectants, which are used to minimize damage to the cells. Commonly used cryoprotectants include: dimethyl sulphoxide (DMSO), glycerol and methanol. There are two components of freezing injury: direct damage from ice crystal formation and secondary damage caused by the increase in concentration of solutes as progressively more ice is formed. Lethal intracellular freezing can be avoided if cooling is slow enough to permit sufficient water to leave the cell during progressive freezing of the extracellular fluid. The rate of freezing varies according to the size of the cells and their water permeability.

A general protocol for microalgae involves cooling the cultures at a rate of -1°C per minute to a holding temperature of around -40°C. The rate of cooling can be controlled using either a programmable freezer or a small portable freezing unit such as a "Mr. Frosty". The samples are then plunged into liquid nitrogen at -196°C and transferred to long term storage. This can be in liquid nitrogen (-196°C), the cold vapor phase above liquid nitrogen (-165°C) or an ultra-cold freezer (-150°C). Storage temperature must not exceed -130°C as this is the threshold for ice re-crystallization, (Mazur, 1966). Cryovials containing specimens should not be left out of the freezer for longer than three minutes to ensure that they do not rise above -130°C.

There are several protocols available for the regeneration of frozen cultures. They all involve removal of the sample from the storage container and rapid transfer into a water-bath set at 37-40°C. The cryovial is gently agitated during thawing until all ice has just melted; this usually takes less than two minutes. The microalgae can then be directly transferred into a flask containing fresh growth medium. Alternatively, the sample may be gently centrifuged and the pellet re-suspended in fresh culture medium. This process is repeated before the culture is transferred to a larger volume of medium. Before returning to normal culturing conditions the sample can be kept in the dark for 24 hours and then in reduced lighting for 24-96 hours. This is important to ensure that cells are not subjected to light levels likely to induce photo-oxidative stress during the recovery phase.

The advantages of cryopreservation of microalgae include (1) it effectively guarantees phenotypic and genotypic stability, (2) it can protect culture against microbial contamination and potential errors arising from handling and labeling during multiple sub-culturing, (3) it requires minimal comparative storage space and (4) it may reduce the cost of long-term maintenance. The disadvantages include (1) initial set-up costs are expensive, (2) storage equipment needs a constant supply of liquid nitrogen, (3) back-up facilities are required in case of mechanical failure, and (4) cultures are not instantly available as post-thaw regeneration times are relatively long.

It must be stressed that there is no universal method for the successful preservation of all microalgal species currently in culture. Different taxonomic groups, and even strains within a species, can exhibit differences in response to the stresses imposed by culture preservation and resuscitation. In general, small easy to culture microalgae tend to be amenable to cryopreservation whereas larger more sensitive species can be problematic.

The ability to cryopreserve samples can be considered as a prerequisite for a culture to be called a "strain". However, as currently only a fraction of algal species are successfully cryopreserved this has not been generally adopted by the phycological community.

Record keeping

When isolating algae, it is essential to keep full records of any strains retained in culture. These should include: an identification number, the name of the species (if known), the collection date, the source of the material, details of the collection site, the method of isolation, the name of the isolator, the culture medium used and any other useful information or special requirements. When maintaining a large number of cultures, it is advisable to store this information in a suitable database with a back-up facility.

References

Andersen, R.A. (2005) Algal Culturing Techniques (Burlington, Massachusetts, U.S.A.: Elsevier Academic Press).

Biddanda, B., Abreu, P.C. and Odebrecht, C. (2006). Algal blooms and bacterial interactions. In Algal Cultures, Analogues of Blooms and Applications, Subba Rao, D. V., ed. (Enfield, New Hampshire, U.S.A.: Science Publishers Inc.), pp. 163-188.

Doucette, G.J., Kodama, M., Franca, S. and Gallacher, S. (1998). Bacterial interactions with Harmful Algal Bloom Species: Bloom Ecology, Toxigenesis, and Cytology. In Physiological Ecology of Harmful Algal Blooms, NATO ASI Series Vol. G41, Anderson, D.M., Cembella, A.D. and Hallegraeff, G.M.; eds. (Heidelburg, Germany: Springer Verlag), pp. 619-647.

Hennenberg, K. J., Fritsche, U. and Herrera, R. (2009). Aquatic Biomass: Sustainable Bio-energy from Algae? Workshop Outcome Report. International Workshop on Sustainable Bioenergy from Algae. Berlin, Germany, 2 November 2009. (Darmstadt: Öko-Institut e.V).

Keller, M.D. Bellows, W.K. and Guillard, R.R.L. (1988). Microwave treatment for sterilization of phytoplankton culture media. J. Exp. Mar. Biol. Ecol. *117*, 279-283.

Lewin, J. (1966). Silicon Metabolism in Diatoms. V. Germanium dioxide, a specific inhibitor of diatom growth. Phycologia *6*, 1-12.

Mahan, K.M., Odom, O.W. and Herrin, D.L. (2005). Controlling fungal contamination in *Chlamydomonas reinhardtii* cultures, BioTechniques *39*, 457-458.

Mazur, P. (1966). Physical and chemical basis of injury in single-celled micro-organisms subjected to freezing and thawing In Cryobiology, Meryman, H. T., ed. (New York, U.S.A.: Academic Press), pp. 213-315.

Meng, X., Yanga, J., Xua, X., Zhanga, L., Nieb, Q., and Xian, M. (2009). Biodiesel production from oleaginous microorganisms. Renewable Energy *34*, 1-5.

Pringsheim, E.G. (1946). Pure Cultures of Algae: Their Preparation and Maintenance, London, U.K.: Cambridge University Press.

Rodrigues, L.H.R., Arenzon, L., Raya-Rodriguez, M.T. and Fontoura, N.F. (2011). Algal density assessed by spectrophotometry: A calibration curve for the unicellular algae *Pseudokirchneriella subcapitata*. J. Environ. Chem. and Ecotoxicol. *3*, 225-228.

Stokner, J. G., Klut, M. E., and Cochan, W. P. (1990). Leaky filters: a warning to aquatic ecologists. Can. J. Fish. Aquat. Sci. *47*, 16–23.

Other Publications of Interest

Flores, E. and Herrero, A. (2014). The Cell Biology of Cyanobacteria. Caister Academic Press, U.K. ISBN: 978-1-908230-38-6

Wilkinson, M.G. (2015). Flow Cytometry in Microbiology: Technology and Applications. Caister Academic Press, U.K. ISBN: 978-1-910190-11-1

Xuefeng, L. (2014). Biofuels: From Microbes to Molecules. Caister Academic Press, U.K. ISBN: 978-1-908230-45-4

3

Phytoplankton Gene Expression

John H. Paul*

College of Marine Science, University of South Florida, 140 Seventh Ave. S., St. Petersburg, FL 33701, U.S.A. jpaul@usf.edu
*Author for correspondence.

Contents

Abstract

Phytoplankton photosynthesis accounts for 50% of the primary production on our planet and forms the base of marine food webs. Phytoplankton can form toxic blooms, take up and release greenhouse gasses, and be responsible for fueling the biological pump. The availability of genomes of several marine phytoplankton including diatoms and expressed sequence tag (EST) libraries of several dinoflagellates have opened the door for understanding gene expression and regulation in this globally important group of photoautotrophs. This review documents early work in phytoplankton gene expression through current studies in marine phytoplanktonic transcriptomes and proteomes. All informatics analyses are hampered by the lack of reliable gene annotations. Progress is being made in gene function identification through the use of endogenous gene silencing mechanisms. Understanding the control of gene networks is in its infancy but promises to be a fertile area of research for the next generation of phytoplankton ocean scientists.

Introduction

A major goal amongst microbial ocean scientists is to understand the *function* of microbial populations in marine environments and how these respond to perturbations on both short and long time scales. There is a real need to understand gene function, not only in marine phytoplankton, but in every biological system on the planet. But what is meant by "gene function"? Tian et al. (2011) points out that the meaning of gene function might depend on one's frame of reference (Tian (2011). A biochemist might describe a gene function in strictly catalytic terms: the enzyme oxidizes a carbonyl group to an organic acid. To a cell physiologist, the participation in a pathway is the gene function. To a clinical researcher, the change in pathogenicity of the microbe caused by gene expression will be important. Thus, every notion of "gene function" inherently has three aspects: molecular function, biological process, and cellular component. Tian et al. pointed out to understand all the actions resulting from gene expression an ontology system covering all major aspects of gene activation was needed. The Gene Ontology (GO) system provides this activity, therefor being better for annotation (Tian et al., 2011).

The GO database is a *relational database* comprising the GO ontologies and the annotations of genes and gene products to terms in the GO. The advantage of housing both the ontologies and annotations in a single database is that powerful queries can be performed over annotations using the ontology.

However, we know now the genome-centric world is a simplistic view of the biosphere, because external genomes can modify gene expression through certain types of symbiotic parasitism. In other situations organisms have outsourced control of their genomes to other microbes in their environment via the process of epigenetics. Temperate and latent viruses are now known to control aspects of the host's gene expression through the action of repressors and transcriptional regulators.

In this review, I discuss the evolution of methods and approaches to the measurement of phytoplankton gene expression. I then address how these approaches are used to measure the regulation of phytoplankton genes involved in C, N, P, and Fe metabolism in the ocean.

The approaches to the measurement of gene expression in the oceans

In a larger sense, gene expression means the detection of the activities encoded in the DNA, or those encoded by the genes identified (open reading frames or ORFs). More narrowly, gene expression refers to the *activation* of the gene from dormant (un-transcribed) to the actively transcribed state (ie. production of a sense transcript). Other notions of gene expression limit the meaning to production of the protein encoded by the ORF in question, while still other concepts of expression are limited to the production of an *active* protein or enzyme. Thus gene expression can encompass transcription, translation, and post-translational processes.

Early gene expression research relied upon detection and sometimes quantification of the target gene mRNA in what was called "northern hybridization" (Kinoshita et al., 1998; Lin and Carpenter, 1998; Pichard, 1993). By this approach, RNA was blotted on charged nylon membranes and hybridized to radiolabeled probes (Paul, 2001). The blots included standards and if these were quantitated one could get accurate concentrations of similar mRNAs in the oceans (Pichard, 1993; Pichard et al., 1996). To verify that RNA and not DNA was being detected, blots were probed with labelled antisense probes to detect transcripts while identical blots were probed with labelled sense probes to correct for non-specific hybridization or contamination of the sample with target DNA (Pichard, 1991). Quantification was accomplished by use of a molecular imager (Paul, 2001). This approach was tedious, usually requiring radioisotopes and production of run-off transcripts and a myriad of other protocols.

An extension of the blot and probe approach to detecting and quantitating gene expression is the microarray approach (Tiquia, 2004). Ward (2008) developed a "phytoarray" that contained oligonucleotides targeted for phytoplanktonic carbon fixation and assimilatory nitrate reduction. This approach successfully detected the shift from diatom C and N assimilation in the early spring bloom to that of haptophytes and flagellates later in the summer (Ward, 2008).

Reverse transcription polymerase chain reaction (RT-PCR) was a sensitive method to detect rare transcripts for which nominal sequence information was available. Le Blanc et al. (1999) developed a semi-quantitative RT-PCR assay encoding a fucoxanthin, chlorophyll a/c-binding protein (FCP) in the centric planktonic diatom *Thalassiosira weissflogii*. With the development of real-time RT-PCR (RT qPCR), detection and quantification of rare transcripts of natural populations has now become a trivial undertaking (Leblanc, 1999). Wawrik et al. (2002) developed a RT qPCR assay for quantitating diatom and pelagophyte ribulose bisphosphate carboxylase/oxygenase large subunit (*rbcL*) transcripts. Although the goal of this research was to design a qPCR assay specific for diatoms only, the pelagophytes could not be resolved from the diatoms based upon *rbcL* sequence data (Wawrik et al., 2002). The desire to have RT qPCR assays for specific ecotypes is often confounded by the reality of sequence conservation or diversity amongst the ecotype groups. For example, although we could easily design a RT qPCR assay for high light *Prochlorococcus*, the low light clade *rbcL* was far too diverse to design a qPCR assay (John et al., 2009). Instead we developed RT qPCR assays for haptophytes (many golden flagellates), form 1A *rbcL* of *Synechococcus,* diatoms/ pelagophytes, and high light *Prochlorococcus* (John et al., 2009; Williamson et al., 2012). Non-specific amplification strategies have been developed to neutralize some of the selectivity of traditional qPCR approaches. mRNA arbitrarily primed reverse transcription-polymerase chain reaction (RAP RT-PCR) is a method to capture unknown mRNAs resulting from a manipulation or a disturbance (Kinoshita et al., 1998). Usually one primer is polydT that initiates cDNA formation of the eukaryotic mRNA pool. Arbitrary primers are then used in the PCR to produce a family of DNAs separated on a 2D gel. This method yields reproducible displays of control and perturbed conditions for cultures.

As mentioned above, transcription does not equal gene expression, even though many biological processes are transcriptionally regulated. A more stringent interpretation of the definition of gene expression is production of the enzyme or structural proteins encoded by that gene. Vergara and co-workers (1998) studied nitrate reductase (NR) expression in the diatom *Thalassiosira weissflogii* (Gru.) Fryxell et Hask. Here they developed antibodies toward the NR as well as detecting NR activity as the penultimate measure of gene expression (Vergara et al., 1998).

Similarly, Paul et al. (2000) measured *rbcL* mRNA, ribulose bisphosphate carboxylase/ oxygenase (RuBisCO) enzyme content, and whole cell carbon fixation in both a marine prymnesiophyte and a marine *Synechococcus* over diel cycles and found all three measurements to co-vary, indicating that transcript abundance was a good proxy for gene expression in this system (Figure 3.1).

Long serial analysis of gene expression (SAGE) has been used to identify *E. huxleyi* genes responsive to nitrogen (N) or phosphorus (P) starvation (Dyhrman et al., 2012).

The genomic approach to understanding gene expression in phytoplankton
As the most successful group of eukaryotic phytoplankton in the modern ocean, diatoms have risen to dominance relatively quickly over the last 100 million years. Recently completed whole diatom genomes have revealed a wealth of information about the evolutionary origins and metabolic adaptations that have led to their ecological success. A major finding is that they have incorporated genes both from their endosymbiotic ancestors and from marine bacteria by horizontal gene transfer. This cacophony of genes encodes novel capacities for metabolic management, for example, allowing the integration of a urea cycle into a photosynthetic cell (Bowler et al., 2013).

Transcriptomics has become a popular way to approach the problem of analyzing phytoplankton gene expression (Ashworth et al., 2013; John et al., 2009; Zielinski, 2014). For eukaryotic phytoplankton, size fractionation of phytoplankton communities or selection of polyA-tailed transcripts (or both) has been used to isolate the eukaryotic fraction of the metatranscriptome. However, this will not enable the capture of mitochondrial or plastid transcripts. Indeed, sometimes plastid transcripts have polyA tails added to identify them for specific degradation (Ward, 2008). After capture on polydT, the RNA is amplified (usually by MessageAmp II, Applied Biosystems). The amplified RNA is then converted to cDNA followed by a second strand synthesis, library construction, and Illumina sequencing.

Proteomics (the study of the sum total of proteins in a cell, also termed the proteome) methods are becoming more widespread in the study of phytoplankton gene expression (Rahman, 2014). The rationale here is to measure production of proteins during gene expression events and not transcripts, which may not be translated into a functional protein. However, the presence of a protein in a host does not mean that protein is catalytically active. This requires a test of its functional capabilities which may not be easily measured.

Case hHistories

Carbon
The study of important gene functions in any organism depends on the completeness of the database for that particular gene and the taxa that might contain it. In 1992, the *rbcL* gene seemed particularly appropriate because of the relative conservation of sequence of

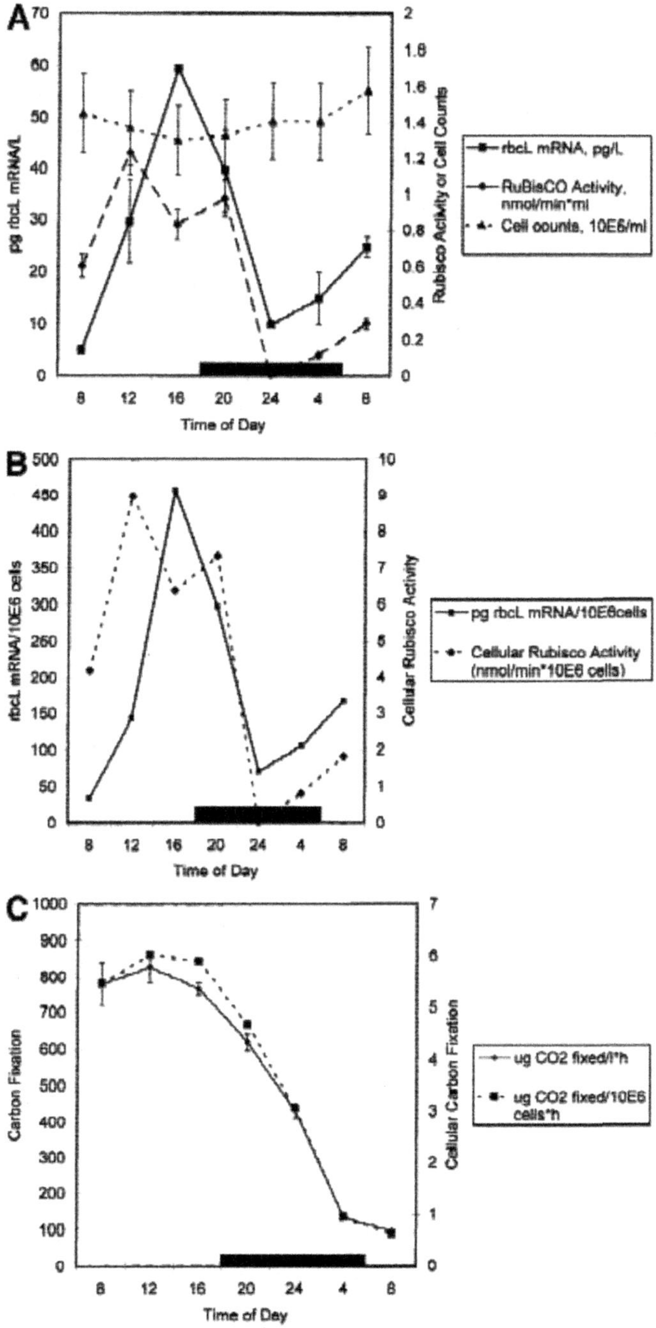

Figure 3.1. Diel studies of the regulation of carbon fixation in *Pavlova gyrans.* (A) *rbcL* transcript abundance, and RuBisCO enzyme activity, and cell abundance over time. (B) Cellular levels of *rbcL* transcript abundance and RuBisCO enzyme activity. (C) Photosynthetic carbon fixation rates (Paul et al., 2000).

this gene, and the significance of c-fixation in marine euphotic communities (Pichard, 1993). This paper used the *rbcL* gene of *Synechococcus* PCC 6301 as a probe of DNA and RNA blots to calculate mRNA/DNA ratios, or what was termed "gene expression per gene dose". At that time it was thought *Synechococcus* PCC 6301 *rbcL* was representative of all phytoplankton but in reality only reflected the Form IB group of RuBisCOs (green algae and higher plants). Therefore, this early attempt to study *rbcL* expression in all phytoplankton would have missed diatoms, haptophytes, and prymnesiophnyes (FormID), *Prochlorococcus* and PE-containing *Synechococcus* (FormIA; (Paul et al., 2000) and dinoflagellates (either FormID or FormII (Tabita, 1999)). Using deck-top incubators, this approach yielded a clear pattern of diel variation in transcription of *rbcL* with the maxima at dawn and early daylight hours and minimal in the dark (Pichard, 1993). Further work using a *Prochlorococcus* strain in culture or natural populations in the Gulf of Mexico found that the diel pattern of *rbcL* transcription was also rhythmically entrained (Pichard et al., 1996). That is, changing natural populations or *Prochlorococcus* in culture from light:dark incubations to continuous illumination resulted in the same entrained pattern of transcription.

The physiological and transcriptomic changes during typical diatom growth in batch culture was studied in *Thalassiosira pseudonana* (Ashworth et al., 2013). Four principal states were investigated: (i) "dawn," following 12 h of darkness; (ii) " dusk," following 12 h of light; (iii) exponential growth and nutrient repletion; and (iv) stationary phase and nutrient depletion. Literally thousands of genes showed increased transcription during the dawn period, including RuBisCO. Reproducible changes in transcription of sensory, signaling, and regulatory functions accompanied the four cell-state transitions, yielding a preliminary road map of how the diatom gene regulatory network operates (Ashworth et al., 2013).

CO_2 fixation in river plumes is a fascinating story of nutrients, light availability, frontal boundaries, and large gradients in salinity. They often dictate phytoplankton C metabolism and modulate phytoplankton production many miles from shore. Our studies have taken us to the Mississippi River Plume (MRP), Orinoco River Plume (ORP), and the Amazon (ARP). The MRP often just loops westward from the Crowsfoot of the Mississippi River delta and results in a zone of hypoxia on the Texas and Louisiana shelf (Rabalais et al., 2002). However when entrained in the Loop Current in the GOM, the plume forms a 10-meter deep feature that is highly visible by water color (Figure 3.2C). Analysis of *rbcL* cDNA libraries indicated that the plume was divided into several phytoplankton domains. Diatoms dominated the most inshore, highly productive waters of the plume. Traveling offshore down the axis of the plume there is a community of PE-containing *Synechococcus,* followed by a second diatom community. The bluewater non-plume stations were dominated by high light *Prochlorococcus* cells (Figure 3.2) (Wawrik and Paul, 2004).

A comparison of the MRP to the Orinoco River Plume (Williamson et al., 2012)indicated that these environments were fundamentally different. The MRP is fueled by nutrients drained from the agriculturally rich drainage basin whereas the Orinoco drains natural mangrove rich lands that results in an order of magnitude less chlorophyll, photosynthesis, nutrient uptake, and total *rbcL* mRNA levels. As shown in other environments, there was a strong diel cycle in *rbcL* mRNA levels and photosynthetic carbon fixation. In both environments, there was a lag between rbcL mRNA and carbon fixation, but the lag was longer in the MRP than the ORP. This lag was primarily in the >2-qμm sized-fraction of the MRP phytoplankton. These results were explained by larger eukaryotic phytoplankton having more mechanisms available to modulate carbon fixation gene expression than the smaller, principally prokaryotic photoautotrophs, being transcriptionally regulated (Williamson et al., 2012).

Equally important in PS carbon fixation as *rbcL* which in part with *rbcS* catalyzes the first step in the Calvin-Bensom-Bassham cycle is the molecular machinery to supply RuBisCO with CO_2, termed the carbon concentrating mechanisms (CCMs). In most prokaryotes that contain RuBisCO the CCM genes are cistronic with *rbcL* and *rbcS* (John et al., 2006). The capacity of phytoplankton to express CCM genes is regulated by environmental factors (Beardall and Giordano, 2002). Some factors such as photon flux can instantaneously change CCM-function without the need to make transcriptional changes. Those environmental parameters that cause changes in the availability of

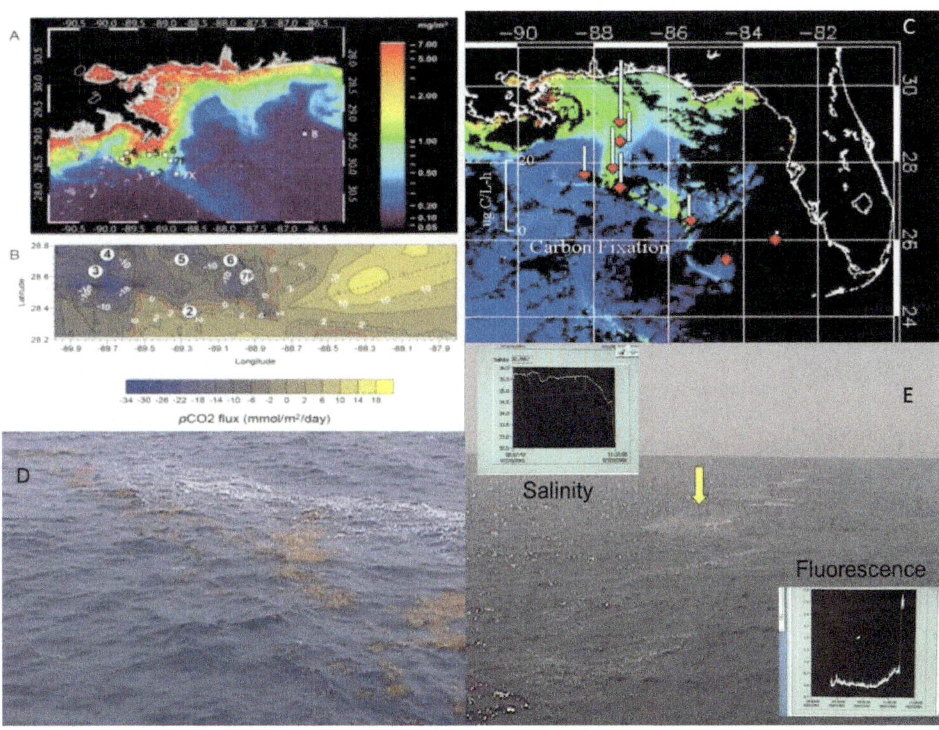

Figure 3.2. (A) MODIS satellite chlorophyll a image of MRP showing station locations. (B) Plot of CO_2 Flux as a function of station location. (C) SeaWiFS chla satellite image of the NE GOM plus carbon fixation rates (Vertical bars) for stations inside and outside the MRP. (D) and (E) The frontal boundary of the MRP in the nearshore (E) and the bluewater plume(D). Note collection of Sargassum at the frontal boundary at the bluewater plume.

dissolved CO_2 in the media surrounding the cells act at the transcriptional level (Beardall and Giordano, 2002).

Another player in the autotrophic carbon acquisition game is the bicarbonate transporters and the carbonic anhydrases. In the bloom-forming cyanoabcterium *Microcystis aeruginosa* there are the ATP-dependent bicarbonate uptake system BCT1 and several carbonic anhydrases (CAs). Strains isolated had the high-flux bicarbonate transporter BicA or the high-affinity bicarbonate transporter SbtA. Growth experiments showed that strains with sbtA performed better at low inorganic carbon (Ci) conditions, whereas strains with bicA performed better at high Ci conditions (Sandrini et al., 2014). Both these genes were on the same operon, and were believed to be regulated by a LysR-type transcriptional regulator.

Carbon concentrating mechanisms and CO_2 fixation pathways are thought to play critical roles in adaptation to increasing pCO_2 resulting from global climate change. *Emiliania huxleyi* is a coccolithiphore of global importance, fixing CO_2 as well as making $CaCO_3$ plates. Acclimation experiments suggest negative effects of warming and acidification on coccolithophore calcification, but the ability of these organisms to adapt to future environmental conditions is not well understood (Benner et al., 2013). Cells exposed to increased inorganic carbon content and calcification rate under warm and acidified conditions compared with ambient conditions, indicating organic carbon content and primary production did not show any change. In contrast to findings from short-term experiments, the results suggest that long-term acclimation or adaptation could change, or even reverse, negative calcification responses in *E. huxleyi* and its feedback to the global carbon cycle. Transcriptome analysis revealed that genes thought to be essential for calcification are not those that are most strongly differentially expressed under long-term exposure to future ocean conditions. Rather, differentially expressed genes observed here represent new targets to study responses to ocean acidification and warming (Benner et al., 2013).

Nitrogen

The interactions between carbon and nitrogen in phytoplankton are complex. Early research efforts, however, often did look at nitrogen independently of carbon or other macronutrients.

Smith et al (1992) used now classical methods to detect diatom response to nutrient flux and increase in PAR one might find in a coastal upwelling. After growing cultures of *S. costatum* under N-limitation and lowlight, shifting into replete nitrate and non-limiting illumination the cells resulted in an increase principally in nitrate uptake and nitrate reductase activity (NRA). Diel fluctuations in NRA were not observed in patterns of change in nitrate reductase protein (Smith et al., 1992).

Regulation of nitrate reductase (NR) was monitored in cultures of *Thalassiosira weissflogii* by monitoring NR protein immunologically as well as nitrate reductase activity. Cells grown on ammonium produced neither NR protein nor activity. Because the RNA synthesis inhibitor actinomycin D and the protein synthesis inhibitor cycloheximide inhibited both NR activity and protein, this argued strongly for regulation of NR at the transcriptional level (Vergara et al., 1998).

The first nitrate transporter (NAT) to be cloned was derived from the diatom *Cylindrotheca fusiformis* Reimann et Lewin. Actually three homologs were discovered, only varying in 18 of 1446 nts. NAT mRNA levels were high in the presence of exogenous nitrate but highly repressed when cells were grown in the presence of NH_4CL or NH_4NO_3. NAT transcript abundance did not occur continuously through the cell cycle, but rather had elevated levels at the beginning of G1 and again at the S-G2 interface. The abundance of silica also influenced the transcription of the NAT genes (Hildebrand and Dahlin, 2000). In *Skeletonema costatum*, phosphate limited cultures had lower levels of the high-affinity nitrate transporter (Nrt2) transcripts than those growing with adequate P (Liu et al., 2013). Compared to the low affinity transporter, the high-affinity Nrt2 transcript levels were *inversely related* to nitrate concentrations below 40 μmol/L (=40 nM; (Liu et al., 2013))

Glutamine synthetase is a key enzyme in nitrogen assimilation as well as a key regulator of cellular N and P metabolism. Transcript abundance of *glnII* (the gene that encodes the GSII isoenzyme), as well as GSII enzyme activity increased in *Skeletonema costatum* (Greville) Cleve cells growing on nitrate but not taking up ammonium (NH_4^+; Takabayashi et al., 2005). Thus, GSII transcript abundance could provide a measure of new production in the oceans.

The impact of N-starvation on the metabolic and lipidomic profiling of 97 low-molecular weight compounds from the primary metabolism and 124 lipid compounds in *Thalassiosira pseudonana* was determined (Bromke et al., 2013). Storage lipids (triglycerides) accumulated as well as TCA cycle intermediates (particularly citric acid which increased 10-fold) in N-starved cultures, whereas amino acid pools were depleted. The increase in TCA intermediates paralleled increases in expression of the TCA cycle genes. Proline, believed to be the major osmoprotectant under N replete conditions was replaced by dimethylsulfoniopropionate under conditions of N-starvation. Therefore, diatoms can alter their metabolic fluxes and metabolite abundances to adapt quickly and efficiently to environmental change (Bromke et al., 2013).

Phosphorus

Not only is phosphorus an essential macronutrient for phytoplankton growth, it can play an important role in harmful algal bloom initiation. Alkaline phosphatase (AP) plays a vital role in organic P utilization. Lin et al.(2012) isolated the AP gene from the HAB *Karenia brevis* and found it to be a cell surface associated protein most closely related to the AP of another toxic dinoflagellate, *Amphidinium carterae*. When *K. brevis* cells were cultivated in low DIP, AP enzyme activity increased 5-6 fold (Lin et al., 2013).

P-starvation did not result in a transcriptional response in *K. brevis* relative to cells grown under P-replete conditions.(Morey J.S.; Monroe, 2011) . However, shift from depleted to replete P conditions resulted in significant changes in 4% of the genes on the *K. brevis* microarray. The earliest responding genes were for pentatricopeptide repeat (PPR) proteins, which increased in expression up to 3-fold by 1 h following nutrient addition. PPR proteins are nuclear encoded proteins involved in chloroplast and mitochondria RNA processing. The nuclear encoded gene transcripts in *K. brevis* all contain a 5' trans-spliced leader (SL) sequence suggestive of regulation by posttranscriptional mechanisms.

A proteomics study of the effect of phosphate deficiency on the marine HAB-forming alga *Aureococcus anophagefferens*. Under times of P-scavaging, relative increases in the protein levels of a phosphate transporter, 5'-nucleotidase, and alkaline phosphatase. P-starvation also resulted in the replacement of phospholipids with sulfolipids (Wurch, 2011).

Iron

Iron has been shown to limit phytoplankton growth, production, or perhaps more correctly gene expression in many portions of the world's oceans. *Prymnesium parvum* was grown under Fe-deplete (0.0025 nM) and Fe-rich (0.05 nM) conditions, and proteomic responses of the organism to Fe conditions were compared (Rahman, 2014). Seven proteins were highly expressed under Fe depleted conditions while one protein was over expressed in Fe-replete conditions as determined by SDS poly acrylamide gel electrophoresis. The proteins were identified by matrix-assisted laser desorption-ionization-time of flight-mass spectrometer (MALDI-TOF-MS) analysis. Under Fe-deplete conditions, *P. parvum* increased the biosynthesis of ATP binding cassette (ABC) transporters, flagellar associated protein (FAP), and phosphoribosyl-aminoimidazole-succinocarboxamide synthase, proteins believed to play a role in Fe-acquisition in phytoplankton. Also apparently upregulated during Fe-depletion were RuBisCO, malate dehydrogenase, and two Fe-independent oxidative stress response proteins, manganese superoxide dismutase (MnSOD) and Serine threonine kinase (STK) (Rahman, 2014). The investigators conclude that *P. parvum* has several gene expression options in responding to Fe depletion as compared to a single gene or operon response.

More than any other group of phytoplankton, large diatoms are particularly susceptible to iron limitation. Iron enrichment experiments typically result in large diatom blooms. A microcosm iron enrichment experiment from the mixed layer of the northeastern Pacific Ocean was performed and transcriptomes generated for enriched and unenriched treatments (Marchetti et al., 2012). In the iron-amended microcosm; transcripts of diatom genes required for synthesis of photosynthesis and chlorophyll components, nitrate assimilation and the urea cycle, and synthesis of carbohydrate storage compounds were significantly overrepresented. Oddly enough, samples receiving supplemental iron had an under-representation of eukaryotic diatom rhodopsins, implying that rodopsins might play a role in photic energy transduction under low iron levels. That is, the iron depleted oceanic phytoplankton might curtail their iron-dependent photophosphorylation (cytochromes, etc.) and derive energy from rhodopsin-like activities (Marchetti et al., 2012).

A comparison of a coastal diatom *Thalassiosira pseudonana*, to its oceanic relative *Thalassiosira oceanica* indicated that the latter was highly tolerant to iron limitation compared to its coastal sibling. Genomic analysis of the two diatoms indicated that the gene encoding ferrodoxin (*petF*) was in the nucleus of *T. pseudonanna* but had been transferred to the nucleus in *T. oceanica* (Lommer et al., 2010). Ferrodoxin can be replaced by flavodoxin during iron limited growth. Presumably the nuclear *petF* location of *T. oceanica* enables this organism to have a competitive advantage in Fe economy during iron limited growth.

Patterns of gene expression

There is value in studying patterns of gene expression by gene cohorts in response to differing environments, environmental perturbations, and time. For example, our interrogation of eukaryotic planktonic meta-transcriptomes through spatial and temporal salinity gradients of the Amazon River Plume (ARP) indicated that stations where surface samples were defined by similar nutrient, sunlight, and salinity regimes resulted in similar patterns of gene transcription (Figure 3.3) (Zielinski, 2014).

Station 10 was a low salinity, nitrate-fueled diatom bloom. Station 10 was a low salinity, nitrate-fueled diatom bloom. Stations 2 and 25 were both diatom diazotroph associations stations (DDAs), with unmeasurable nitrate and an intermediate salinity. These two stations were sampled nearly a month apart, yet retained a similar biogeochemical and metatranscriptomic identity (Figure 3.4) (Zielinski, 2014).

Analysis of 130 000 ESTs from *Phaeodactylum tricornutum* cells grown in 16 different conditions have resulted in similar expression profiles and identified transcripts differentially expressed in response to specific treatments (Maheswari et al., 2010).

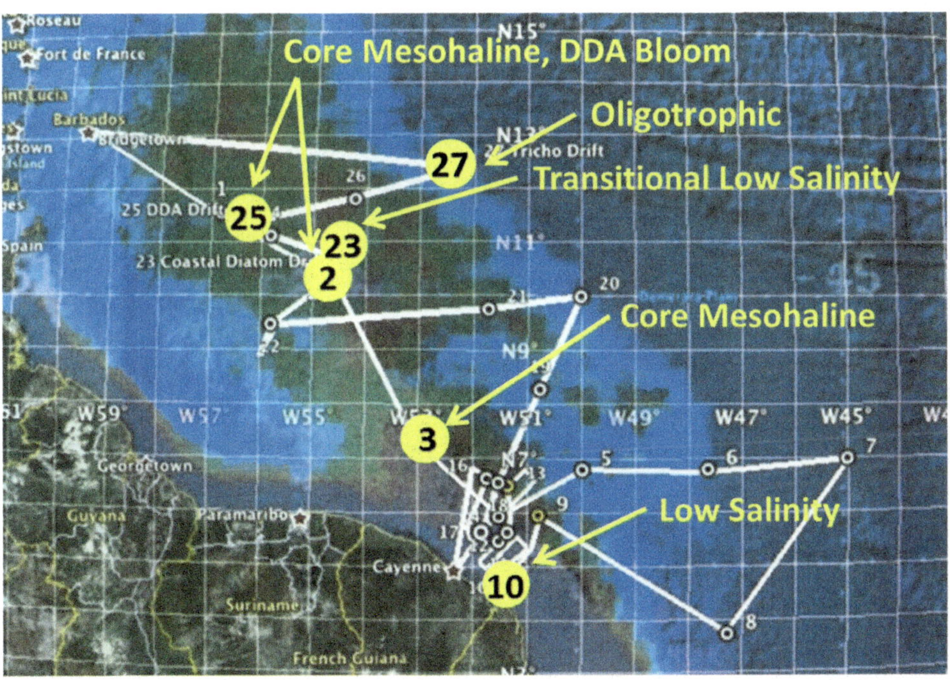

Figure 3.3. Cruise track of the R/V Knorr in the ARP 2010. Meta-Transcriptomic samplings occurred at stations indicated by yellow circles.

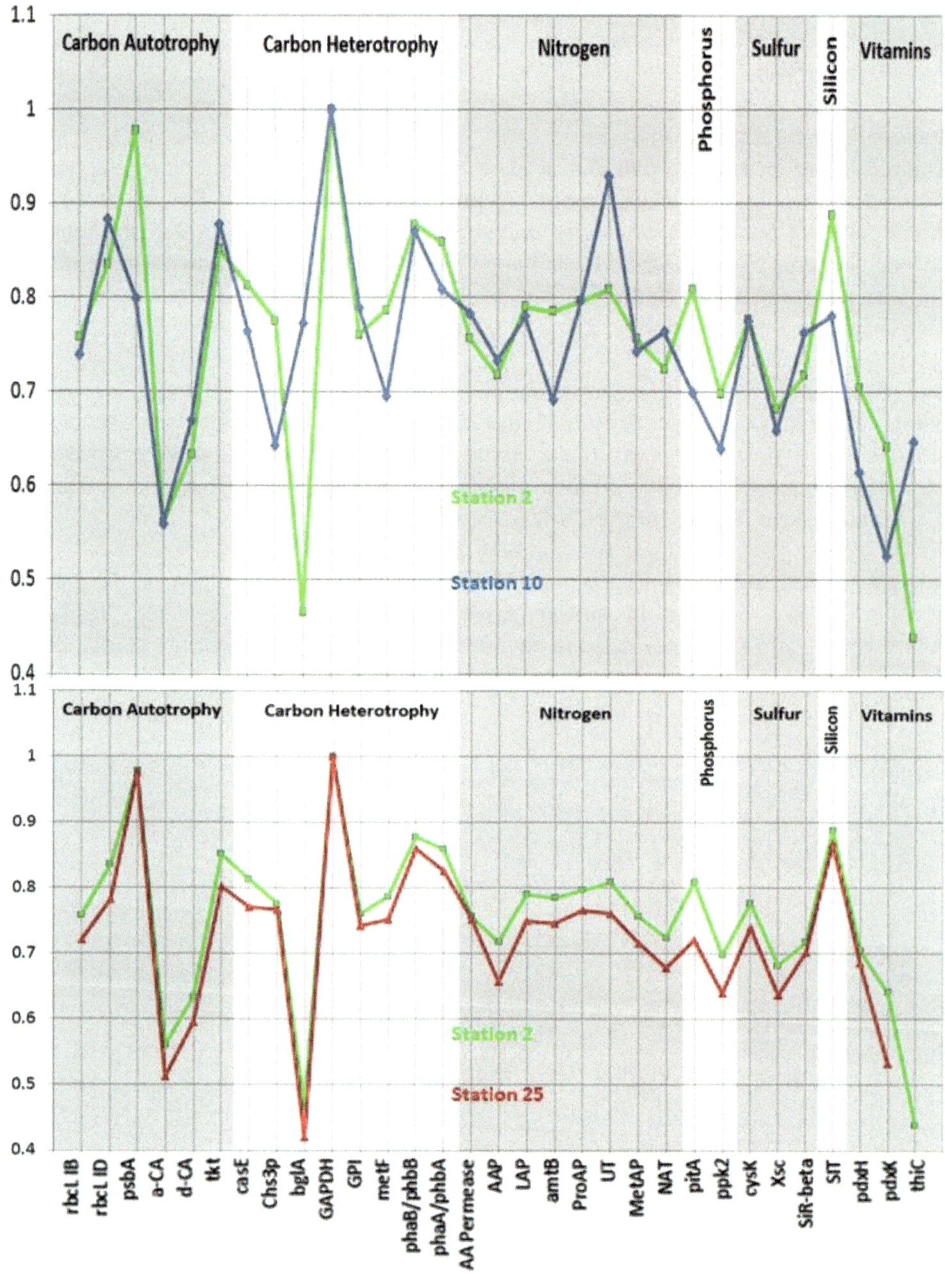

Figure 3.4. Analysis of patterns of gene expression (>2 µm particle size) of polyA tailed transcripts from stations 2 and 10 (top panel) and 2 and 25 (station locations in Figure 3.3). Thirty-one genes (X-axis) were BLASTED against each station's transcriptomes. Then, each transcriptome was normalized to the transcript abundance for glyceraldehyde phosphate dehydrogenase (GAPDH), a common housekeeping gene in both autotrophs and heterotrophs (which yields the Y-axis values, ranging from 0 to 1). The nearly identical patterns of expression between Station 2 and 25 and the disparate patterns between 2 and 10) is noteworthy.

Functional annotation of these transcripts provides insights into expression patterns of genes involved in various metabolic and regulatory pathways and into the roles of novel genes with unknown functions. Specific growth conditions resulted in greater gene diversity, known gene product functions, and over-representation of rare and/or unknown transcripts. When comparative expression analysis was done with *Thalassiosira pseudonana*, it was possible to identify unique diatom genes that were specifically regulated to a particular set of conditions. This common transcriptomic response enabled studies of gene function, genome annotation, and the genomic basis of species diversity and phenomics (Maheswari et al., 2010).

Omics and the future of phytoplankton gene expression

Years ago the study of the regulation of production of an enzyme or several enzymes in any biological system was called "enzymology" (the term "proteomics" had yet to be invented). When such processes were studied in the field investigators were armed with one or two assays and correlated gene-function (a term for expression) to one or several processes or environmental variables. Things were simple. Today proteomics papers are becoming commonplace that look for the over-representation of transcripts (what was previously termed upregulation) in response to an environmental perturbation (Fe limitation, increase in pCO_2, etc.) and hundreds of transcripts arise as potentially important candidates in the regulation of a process. The conclusions? The organism has multiple genes, pathways, and processes responding to the stimuli (Dyhrman et al., 2012). Are all processes needed to elicit the gene function of concern or are the secondary processes fortuitously co-regulated?

The core problem in making sense of gene expression in phytoplankton is the lack of functional annotation research to truly identify the tremendous number of highly expressed or repressed genes that are either responding to a perturbation or resulting in a different phenome (or both). Although bioinformatics has become extremely sophisticated through PFAM, hidden Markov models, and like analysis in the attempt to correctly annotate gene function, there is still a tremendous need to do what was previously called "genetics". That is, through gene knockout/knockdown analysis to pinpoint gene function for an orfan, a conserved hypothetical ORF, and the like. However, genetic systems or means of site directed mutagenesis and manipulation are lacking in most environmentally significant or relevant taxa. Using the diatom *Phaeodactylum tricornutum* it has been demonstrated that gene silencing mechanisms exist to trigger knockdowns and knockouts to identify gene function (De Riso et al., 2009). Using either inverted repeats or antisense constructs of selected target genes it has been possible to silence a GUS reporter gene as well as the knockdown of endogenous phytochrome (DPH1) and cryptochrome (CPF1) genes. Initial molecular analyses reveal that targeted downregulation likely occurs through transcriptional and post-transcriptional gene silencing mechanisms (De Riso et al., 2009). This approach, albeit tedious, blazes the trail for reverse genetics analysis to understand the biology and ecology of diatoms.

Omics data can be organized along the flow of biological information: genomics, transcriptomics, proteomics, metabolomics, and phenomics. Briefly, genomics is the total genome sequence of an organism, including ORFs, regulatory elements, and non-coding RNAs. Transcriptomics is the sum total of all the transcripts in the organism. Proteomics is the sum total of all proteins in the cell, and metabolomics refers to the intracellular pools of small mw metabolites. The sum total of all observable traits in an organism is the field of "phenomics", also termed the sum of all other omics (Tian, 2011).

For a single organism, this approach makes sense, even though the integration of all omics in one database has yet to occur for any organism. As we see the various "meta-omics" (ie meta-genomics, meta-transcriptomics, meta-proteomics), one must consider yet another layer of organization, meta-phenomics. Meta-phenomics is the sumtotal of all observable phenotypes in a system. The interaction of phenotypes in an environment is a definition of ecology. On a global scale, could the sum total of the various ecological systems on this planet be called the Eco-nome? If so, the term for the field would be "*Economics*", ironically sharing the name for "*A social science that studies how individuals, governments, firms and nations make choices on allocating scarce resources to satisfy their unlimited wants.*"

More correctly, the aggregate of all ecological interactions on the planet would be the "Ecolome".

Web resources

Transcriptomics

The Moore Foundation initiated a Marine Microbial Eukaryote Transcriptome Sequencing Project (http://marinemicroeukaryotes.org). As transcriptomes emerge from this initiative, the entire marine eukaryotic phytoplankton database will exponentially grow.

Proteomics

ExPASy bids itself as a Bioinformatics Resource Portal and contains links to a myriad of databases and web tools. http://www.expasy.org/proteomics

Metabolomics

Perhaps the largest collection of plant and microbial metabolomics web resources:
http://www.bmrb.wisc.edu/metabolomics/external_metab_links.html

Phenomics

An interesting initiative of Avotol is Next Generation Phenomics for the Tree of Life. A multi-institutional collaboration to increase the use of phenomic data for phylogenetic tree-building by using: Crowd Sourcing, Natural Language Processing, and Computer Vision. http://avatol.org/ngp/

Acknowledgements

This research was supported by the Gordon and Betty Moore Foundation as part of the River Ocean Continuum of the Amazon (ROCA) project (Grants GBMF #2293 and #2928). Special thanks to Brian Zielinski and Patricia Yager for helpful discussions.

References

Ashworth, J., Coesel, S., Lee, A., Armbrust, E.V., Orellana, M.V., and Baliga, N.S. (2013). Genome-wide diel growth state transitions in the diatom *Thalassiosira pseudonana*. Proc. Natl. Acad. Sci. U.S.A. *110*, 7518-7523.

Beardall, J., and Giordano, M. (2002). Ecological implications of microalgal and cyanobacterial CO_2 concentrating mechanisms, and their regulation. Funct. Plant Biol. *29*, 335-347.

Benner, I., Diner, R.E., Lefebvre, S.C., Li, D., Komada, T., Carpenter, E.J., and Stillman, J.H. (2013). *Emiliania huxleyi* increases calcification but not expression of calcification-related genes in long-term exposure to elevated temperature and pCO(2). Philos Trans R Soc B-Biol Sci *368*, 17.

Bowler, C.A., Karger, D.A., Patzke, E.B., and Stubbs, J.R. (2013). Using instructional design principles to create training for a large transfusion service's Document Control Conversion. Transfusion *53*, 248A-248A.

Bromke, M.A., Giavalisco, P., Willmitzer, L., and Hesse, H. (2013). Metabolic analysis of adaptation to short-term changes in culture conditions of the marine dDiatom *Thalassiosira pseudonana*. PLoS One *8*, 11.

De Riso, V., Raniello, R., Maumus, F., Rogato, A., Bowler, C., and Falciatore, A. (2009). Gene silencing in the marine diatom *Phaeodactylum tricornutum*. Nucleic Acids Research *37*, E96-E96.

Dyhrman, S.T., Jenkins, B.D., Rynearson, T.A., Saito, M.A., Mercier, M.L., Alexander, H., Whitney, L.P., Drzewianowski, A., Bulygin, V.V., Bertrand, E.M., *et al.* (2012). The transcriptome and proteome of the diatom *Thalassiosira pseudonana* reveal a diverse phosphorus stress response. PLoS One *7*, 10.

Hildebrand, M., and Dahlin, K. (2000). Nitrate transporter genes from the diatom *Cylindrotheca fusiformis* (Bacillariophyceae): mRNA levels controlled by nitrogen source and by the cell cycle. Journal of Phycology *36*, 702-713.

John, D.E., Wawrik, B., Tabita, F.R., and Paul, J.H. (2006). Gene diversity and organization in rbcL-containing genome fragments from uncultivated *Synechococcus* in the Gulf of Mexico. Marine Ecology Progress Series *316*, 23-33.

John, D.E., Zielinski, B.L., and Paul, J.H. (2009). Creation of a pilot metatranscriptome library from eukaryotic plankton of a eutrophic bay (Tampa Bay, Florida). Limnol Oceanogr Meth *7*, 249-259.

Kinoshita, S., Kikuchi, K., Yamada, H., Hara, T., Itoh, Y., Kinoshita, H., and Watabe, S. (1998). Differential expression of mRNAs in the marine diatom *Chaetoceros compressum* exposed to high temperatures. Fisheries Science *64*, 831-835.

Leblanc, C.F., A ; Watanabe, M; Bowler, C (1999). Semi-quantitative RT-PCR analysis of photoregulated geneexpression in marine diatoms. Plant Molecular Biology *40*, 1031-1034.

Lin, H.Y., Shih, C.Y., Liu, H.C., Chang, J., Chen, Y.L., Chen, Y.R., Lin, H.T., Chang, Y.Y., Hsu, C.H., and Lin, H.J. (2013). Identification and characterization of an extracellular alkaline phosphatase in the marine diatom *Phaeodactylum tricornutum*. Mar. Biotechnol. *15*, 425-436.

Lin, S.J., and Carpenter, E.J. (1998). Identification and preliminary characterization of PCNA gene in the marine phytoplankton *Dunaliella tertiolecta* and *Isochrysis galbana*. Molecular Marine Biology and Biotechnology *7*, 62-71.

Liu, Y., Song, X.X., Han, X.T., and Yu, Z.M. (2013). Influences of external nutrient conditions on the transcript levels of a nitrate transporter gene in *Skeletonema costatum*. Acta Oceanol Sin *32*, 82-88.

Maheswari, U., Jabbari, K., Petit, J.L., Porcel, B.M., Allen, A.E., Cadoret, J.P., De Martino, A., Heijde, M., Kaas, R., La Roche, J., *et al.* (2010). Digital expression profiling of novel diatom transcripts provides insight into their biological functions. Genome Biol *11*, 19.

Marchetti, A., Schruth, D.M., Durkin, C.A., Parker, M.S., Kodner, R.B., Berthiaume, C.T., Morales, R., Allen, A.E., and Armbrust, E.V. (2012). Comparative metatranscriptomics identifies molecular bases for the physiological responses of phytoplankton to varying iron availability. Proc. Natl. Acad. Sci. U.S.A. *109*, E317-E325.

Morey J.S.; Monroe, E.K., AL ; Beal, M ; Johnson, JG; Hitchcock, GL ; Van Dolah, FM (2011). Transcriptomic response of the red tide dinoflagellate, *Karenia brevis*, to nitrogen and phosphorus depletion and addition. BMC Genomics *12*.

Paul, J.H. (2001). Gene expression by mRNA analysis. Methods in Microbiology, Vol 30 *30*, 395-408.

Paul, J.H., Kang, J.B., and Tabita, F.R. (2000). Diel patterns of regulation of rbcL transcription in a cyanobacterium and a prymnesiophyte. Mar Biotechnol *2*, 429-436.

Pichard, S.F., ME ; Paul, JH (1993). Ribulose-bisphosphate carboxylase gene-expression in subtropical marine-phytoplanktonpopulations. Marine Ecology Progress Series *101*, 55-65.

Pichard, S.L., Campbell, L., Kang, J.B., Tabita, F.R., and Paul, J.H. (1996). Regulation of ribulose bisphosphate carboxylase gene expression in natural phytoplankton communities .1. Diel rhythms. Marine Ecology Progress Series *139*, 257-265.

Pichard, S.L.a.J.H.P. (1991). . Detection of gene expression in genetically engineered microorganisms and natural phytoplankton populations in the marine environment by mRNA analysis. . Appl Environ Microbiol *57*, 1721-1727.

Rabalais, N.N., Turner, R.E., and Wiseman, W.J. (2002). Gulf of Mexico hypoxia, aka "The dead zone". Annu Rev Ecol Syst *33*, 235-263.

Rahman, M.R., MA ; Maki, T ; Nishiuchi, T; Asano, T ; Hasegawa, H (2014). A marine phytoplankton (*Prymnesium parvum*) up-regulates ABC transporters and several other proteins to acclimatize with Fe-limitation. J chemosphere *95*, 213-219.

Sandrini, G., Matthijs, H.C.P., Verspagen, J.M.H., Muyzer, G., and Huisman, J. (2014). Genetic diversity of inorganic carbon uptake systems causes variation in CO_2 response of the cyanobacterium *Microcystis*. Isme J *8*, 589-600.

Smith, G.J., Zimmerman, R.C., and Alberte, R.S. (1992). Molecular and physiological-responses of diatoms to variable levels of irradiance and nitrogen availability - growth of *Skeletonema-Costatum* in simulated upwelling conditions. Limnol Oceanogr *37*, 989-1007.

Tabita, F.R. (1999). Microbial ribulose 1,5-bisphosphate carboxylase/oxygenase: A different perspective. Photosynth Res *60*, 1-28.

Takabayashi, M., Wilkerson, F.P., and Robertson, D. (2005). Response of glutamine synthetase gene transcription and enzyme activity to external nitrogen sources in the diatom *Skeletonema costatum* (*Bacillariophyceae*). Journal of Phycology *41*, 84-94.

Tian, W.D., X.; Zhou, W.; Ren R. (2011). Predicting gene function using omics data:
from data preparation to data integration. In Protein Function Prediction for Omics Era,, D. Kihara, ed., pp. 215-242.

Tiquia, S.W., LY; Chong, SC; Passovets, S; Xu, D (Xu, D); Xu, Y; Zhou, JZ (2004). Evaluation of 50-mer oligonucleotide arrays for detecting microbial populations in environmental samples. Biotechniques *36*, 664-.

Vergara, J.J., Berges, J.A., and Falkowski, P.G. (1998). Diel periodicity of nitrate reductase activity and protein levels in the marine diatom *Thalassiosira weissflogii* (*Bacillariophyceae*). Journal of Phycology *34*, 952-961.

Ward, B.B. (2008). Phytoplankton community composition and gene expression of functional genes involved in carbon and nitrogen assimilation. Journal of Phycology *44*, 1490-1503.

Wawrik, B., and Paul, J.H. (2004). Phytoplankton community structure and productivity along the axis of the Mississippi River plume in oligotrophic Gulf of Mexico waters. Aquat Microb Ecol *35*, 185-196.

Wawrik, B., Paul, J.H., and Tabita, F.R. (2002). Real-time PCR quantification of rbcL (ribulose-1,5-bisphosphate carboxylase/oxygenase) mRNA in diatoms and pelagophytes. Applied and Environmental Microbiology *68*, 3771-3779.

Williamson, S.J., Allen, L.Z., Lorenzi, H.A., Fadrosh, D.W., Brami, D., Thiagarajan, M., McCrow, J.P., Tovchigrechko, A., Yooseph, S., and Venter, J.C. (2012). Metagenomic exploration of viruses throughout the Indian Ocean. PLoS ONE *7*, e42047.

Wurch, L.L.B., E.M. Saito, M.A. ; Van Mooy, B.A.S.; Dyhrman, S.T. (2011). Proteome changes driven by phosphorus deficiency and recovery in the brown tide-forming alga *Aureococcus anophagefferens*. PLoS One *6*.

Zielinski, B.L.B.Z., Allen, A; Carpenter, EJ; Coles, V: Crump, B; Mary Daugherty, Rachel Foster, Joachim Goes, Helga Gomes, John McCrow, Joseph Montoya, Mary Ann Moran, Brandon Satinsky, Shalabh Sharma, Christa Smith, Sarah Weber, Tish Yager, John H Paul (2014). Patterns of microeukaryotic gene expression derived from metatranscriptome analysis parallels biogeochemical measurements in the Amazon river plume. in review.

Other Publications of Interest

Flores, E. and Herrero, A. (2014). The Cell Biology of Cyanobacteria. Caister Academic Press, U.K. ISBN: 978-1-908230-38-6

Herold, K.E. and Rasooly, A. (2009). Lab-on-a-Chip Technology: Fabrication and Microfluidics. Caister Academic Press, U.K. ISBN: 978-1-904455-46-2

Herold, K.E. and Rasooly, A. (2009) Lab-on-a-Chip Technology: Biomolecular Separation and Analysis. Caister Academic Press, U.K. ISBN: 978-1-904455-47-9

Wilkinson, M.G. (2015). Flow Cytometry in Microbiology: Technology and Applications. Caister Academic Press, U.K. ISBN: 978-1-910190-11-1

Xuefeng, L. (2014). Biofuels: From Microbes to Molecules. Caister Academic Press, U.K. ISBN: 978-1-908230-45-4

4

Marine Microalgae and Harmful Algal Blooms: A European Perspective

Johanne Arff[1]*and Belén Martìn Miguez[2]*

[1] SINTEF Fisheries and Aquaculture, Box. 4762 Sluppen, N7465 Trondheim, Norway. Johanne.Arff@sintef.no
[2]CETMAR, Centro Tecnológico del Mar-Fundación, Eduardo Cabello s/n, 36208 Bouzas-Vigo, Spain. bmartin@cetmar.org
*Both authors for correspondence.

Contents

Abstract

Harmful Algal Blooms (HABs) are a global phenomenon impacting both wild life and human health; especially the effects of toxin-producing microalgae can be severe, and monitoring programs have been established as a risk reducing measure. However, recent research has demonstrated that HAB species that previously were believed to have a restricted geographical distribution have spread to new areas. Moreover, monitoring programs on seafood safety are often designed to detect harmful algae and biotoxins already known in an area; thus the establishment of new HAB species might not be detected until the effects of a HAB event is evident (i.e. fish killings or human seafood poisoning). Furthermore, an increased understanding of the links between oceanographic conditions and the development of HABs, as well as their toxic properties will be important in future monitoring of HAB species.

Introduction

Algal blooms are a natural phenomenon being an important part of the marine food web, and thus the basis for marine production and the human exploitation of marine resources as a food provider. However some algal blooms are considered to be harmful to humans, marine mammals, birds, fish or the marine ecosystem as a total either due to the effects from high biomass or the production of biotoxins. The natural toxins produced by some microalgae are, however, of the uttermost concern due to their ability to accumulate in seafood; hence posing a risk towards humans consuming contaminated seafood. On a yearly basis, almost 2000 cases of food poisoning from consumption of toxic fish and shellfish are reported worldwide with about 15% having a fatal outcome (Hallegraeff, 2003). Moreover, due to their fish-killing ability blooms of ichthyotoxic microalgae pose a threat not only towards a growing aquaculture industry, but also to wild fish stocks and the fisheries.

HAB species are found among the diatoms, dictyochophycea, dinoflagellates, haptophytes, raphidophycea and the cyanobacteria, with dinoflagellates making up the majority of the toxin producing microalgae. Due to their potential risk towards animal life and human health, monitoring programs have been established which give valuable insight on the geographic distribution of HAB species. Here, we describe some important HAB species, focusing on their European distribution.

Getting insight into the oceanographic settings that may favour the onset of HABs is essential to understand their dynamics and eventually better manage the marine ecosystems affected by them. In this chapter, we will describe the interactions between oceanographic conditions and HABs following the approach put forward by the international GEOHAB (Global Ecology and Oceanography of Harmful Algal Blooms) program, developed around four types of ecosystems: (1) fjords and coastal embayments, (2) upwelling systems, (3) eutrophic systems and (4) stratified systems.

Harmful marine microalgae

Diatoms

Diatoms are one of the largest algal groups known, considered to encompass 10 000 to 12 000 species (Fryxell and Hasle, 2003). Recently Mann and Vanormelingen (2013) have estimated the number of diatom species to be at least 30 000, and they suggest as much as a diversity of 100 000 different diatoms as a probable number of species. Diatoms are an ecological important algal group with an estimated contribution of 40-45 % of the oceans primary production.

Any bloom of diatoms may be considered as potentially harmful due to ecological effects such as post-bloom anoxic water, production of oil or mucus, bitter taste in shellfish or toxin production (Fryxell and Hasle, 2003). Even if there is a vast number of diatoms, there are at present only 14 known toxin-producing species (Moestrup et al., 2013) constricted to the genera *Pseudo-nitzschia* (12 species), *Halamphora* (1 species; syn. *Amphora*) and *Nitzschia* (1 species) all pennate diatoms. All toxic diatoms are identified as producers of the neurotoxin domoic acid (DA); the toxin responsible for amnesic shellfish poisoning (ASP). The first incident of ASP was reported from Canada in November 1987 with more than 100 cases of illness, three having a fatal outcome, after consumption of toxic mussels. DA was detected and *Pseudo-nitzschia multiseries* was identified as the source organism (Bates et al., 1989). This incident lead to an increasing concern regarding ASP outbreaks and several countries included DA and *Pseudo-nitzschia* in their monitoring programs on seafood safety resulting in the identification of other DA producing *Pseudo-nitzschia* species. DA is also recognized as the causative of mortalities of seabirds and sea mammals in California and Mexico feeding on contaminated anchovies or mackerel; *Pseudo-nitzschia australis* being the source organism in all cases (Lefebrve and Robertson, 2010). *Pseudo-nitzschia* is also common in Europe with reported blooms from Norway, Denmark, U.K., Ireland, Spain and Portugal (Smayda, 2006).

Accumulation of DA in different shellfish species was studied after a DA outbreak in Irish waters in 1999 (James et al., 2005). This study demonstrated that king scallops (*Pecten maximus*) are a more potent vector than blue mussels (*Mytlius edulis*), oysters (*Crassostrea edulis*) and razor clams (*Ensis siliqua*). Moreover, DA accumulates differently in different tissue in *P. maximus* with highest concentrations, and well above the regulatory limit of 20 µg DA/g in hepatopancreas with values below the regulatory limit in the abductor muscle (James et al., 2005). There are several studies on *Pseudo-nitzschia* spp. and DA in European waters; especially with respect to nutrients as a bloom regulator and spatial distribution of DA in bivalva. Bogan et al. (2007a, 2007b) demonstrated that the DA content in scallops from different harvesting grounds within relatively limited geographical areas (i.e. around Isle of Man and southern parts of Ireland) differed significantly, hence posing a risk with respect to the design of monitoring programs and the regulation of harvesting areas. Furthermore, they discussed how different oceanographic conditions influencing the nutrient dynamics might explain the observed spatial differences of DA in shellfish. Another interesting feature with *Pseudo-nitzschia* spp. that may explain why higher DA values are detected in scallops than in other shellfish are the accumulation in thin layers, sometimes in depths below ten metres (Rines et al., 2010), which is well below the depths where mussels, oysters and clams are produced or harvested from, while scallops can have a vertical distribution as deep as 100 m (Hayward et al., 1996) and as such are more exposed for *Pseudo-nitzschia.* Furthermore, the sedimentation of *Pseudo-nitzschia* out of the productive zone might be another mechanism explaining the higher concentrations of DA in scallops than in other shellfish. However, to the authors' knowledge, the European monitoring programs on *Pseudo-nitzschia* and DA have been successful with no reports on human intoxications due to the consumption of DA contaminated shellfish so far in Europe.

There are several reports on mass occurrences of non-toxic diatoms and fish mortality; most of them related to aquaculture (i.e. caged fish are unable to escape unfavourable environmental conditions). Typically, diatoms harm the fish by either clogging or damaging the gills, thus reducing the oxygen uptake through them. As a reaction to the diatoms, the gill tissue produces mucus. Mortality is not caused by the diatoms itself but as a result of reduced oxygen uptake (hypoxia) and carbon dioxide excretion (hypercapnia) through the gill tissue (Rensel, 1993). Species belonging to the genera *Cerataulina, Chaetoceros, Corethron, Leptocylindrus, Rhizosolenia, Skeletonema,* and *Thalassiosira* have induced mortality in farmed fish (Smayda, 2006). Especially the large *Chaetoceros* belonging to the subgenus *Phaeoceros* have protruding spines on their setae which can loosen from the algal cell and penetrate the gills resulting in bleeding. As refered by Smayda (2006) Chilean aquaculture has had losses due to blooms of *Chaetoceros* and *Leptocylindrus;* while *Corethron* and *Chaetoceros* have killed cultured salmonides in Canada (Yang and Albright, 1994; Speare et al., 1989). *Chaetoceros* spp., *Skeletonema* and *Thalassiosira* spp. have been reported as the causative of mortalities in Scottish fish farms (Treasurer et al., 2003; Kent et al., 1995). Smayda (2006) also refers to two episodes in New Zealand where *Cerataulina pelagica* has killed fish respectively by clogging the gills and being an effect of post-bloom anoxia.

Furthermore, there are also reports on diatoms impacting European fisheries. In 1977, *Coscinodiscus wailesii* was first recorded in European waters; soon after local fishermen reported on excessive mucus clogging the gear (Boalch, 1987). Moreover, Boalch reports that *C. wailesii* was demonstrated to produce mucus in laboratory cultures, and *C. wailesii* was recorded in samples collected in areas affected by mucus. This problem only lasted a few years, and Boalch suggested that this might be attributed to the observation of a change in dominating season for *C. wailesii;* i.e. from summer to winter and spring plankton. According to Smayda (2006) *C. wailesii* has also been a problem for Japanese nori-production; competing for nutrients and hence reducing the yield. Smayda also refers to bird mortality in connection with a bloom of *C. concinnus* in the North Sea.

Dinoflagellates

There are 2 377 described dinoflagellate species belonging to 259 genera (Gómez et al., 2012) and of these are nearly 80 species listed as toxin-producing species (Moestrup et al. 2013). The majority of known toxin-producing microalgae belong to the dinoflagellates and until the early 1980s only dinoflagellates were identified as HAB species. Dinoflagellates produce a wide range of neurotoxins being responsible for ciguatera fish poisoning (CFP), neurotoxic shellfish poisoning (NSP), diarrheic shellfish poisoning (DSP), paralytic shellfish poisoning (PSP), and other harmful events such as respirational irritation in humans and animal mortalities.

Dinophysis produce lipophilic toxins; i.e. okadaic acid (OA) and its derivates dinophysistoxins (DTXs), and pectenotoxins (PTX), which are associated with DSP outbreaks worldwide. The first reports on gastrointestinal outbreaks, which later have been set in connection with *Dinophysis* and diarrheic shellfish poisoning (DSP), were from the Netherlands and Chile in the 1970s (Reguera et al., 2012 and references therein). It was, however, not until after the outbreaks in Japan in 1976 and 1977 the responsible toxin, OA, was isolated and the causative organism, *D. fortii,* was identified.

In the 1980s, Denmark, France, Germany, Ireland, Italy, Norway, Portugal, Spain and Sweden experienced DSP events with several human intoxications after the consumption of toxic shellfish (Smayda 2006). *Dinophysis* and DSP is now a recurring phenomenon in European waters with concentrations above regulatory levels, and are of special concern for the shellfish industry resulting in closure of harvesting areas. Moreover, benthic species belonging to *Prorocentrum* have also been demonstrated to produce dinophysistoxins, and shellfish feeding on resuspended toxic *Prorocentrum* might also be a source to DSP (Smayda 2006).

Dinophysis consists of over 100 species, and at present twelve species are listed as toxic, including ten *Dinophysis* spp and two *Phalacroma* spp. (Reguera et al., 2012). Furthermore, they have sorted the *Dinophysis* species responsible for DSP events by latitude; with *D. acuminata, D. acuta* and *D. norvegica* in boreal seas*; D. acuminata/D. sacculus/D. ovum, D. acuta, D. fortii* and *D. caudata* in temperate seas; and finally *D. caudata* and *D. miles* in tropical seas. The *Dinophysis* species concept is, however, complicated with different morphotypes in different geographic regions and the misidentification of the polymorphic life cycle stages of a species as different species is common.

Yessotoxins (YTXs) are another group of lipophilic toxins that have been demonstrated to accumulate in shellfish; with *Protoceratium reticulatum, Lingulodinium polyedrum* and *Gonyaulax spinifera* as recognized causative species (Paz et al., 2008). YTX was first isolated in scallops from Japan (i.e. *Patinopecten yessoensis*) in 1986, and has since then been detected in mussels from Norway, Italy, Spain, Russia, New Zealand, and Chile (Paz et al., 2008 and references therein). According to Paz et al. (2008), there are no reports on human intoxications due to YTX. Yessotoxins have, however, been demonstrated to be cytotoxic and are included in European shellfish monitoring programs.

Saxitoxin (STX) and derivates are responsible for paralytic shellfish poisoning; with *Gymnodinium catenatum, Pyrodinium bahamanese* var. *compressum* and *Alexandrium* spp. as identified source organisms. *Gymnodinium catenatum* is the only STX-producing athecate dinoflagellate. The distribution of *G. catenatum* has recently been reviewed by Hallegraeff et al. (2012); and according to their findings *G. catenatum* is more wide-spread in warm-temperate and sub-tropical waters than previously believed with records in Europe, North and South America, Asia and Austrailasia (Guiry and Guiry, 2013). In Europe *G. catenatum* is recorded in the Mediterranean and along the Atlantic coast off the Iberian Peninsula; furthermore the first record of *G. catenatum* and PSP is from Spain in 1976 (see Hallegraeff et al., 2012). HABs of *G. catenatum* and subsequent closure of harvesting areas have since then become a recurring challenge for the Spanish shellfish industry located in Galician Rias.

The first report of a PSP outbreak after consumption of shellfish contaminated by *Pyrodinium bahamense* was from Papua New Guinea in 1975 when several locals became ill and were hospitalized, some cases with a fatal outcome (Worth et al., 1975). Two varieties are described for *P. bahamense*, the toxic *P. bahamense* var. *compressum* and the non-toxic *P. bahamense* var. *bahamense,* having different geographical distribution. The toxic variety is recorded from south-east Asian waters, the Indian Ocean and the Pacific coast of Central America, while the non-toxic is described from the tropical West Atlantic (Taylor et al., 2003). Usup et al. (2012), however, refer to recent studies from Florida where the non-toxic *P. bahamense* var. *bahamense* has been demonstrated to produce STX, thus being the causative of PSP events after consumption of STX contaminated puffer fish.

Alexandrium is distributed all over the world with 33 recognized species (Mackenzie and Todd, 2002; Montresor et al., 2004) dived into two subgenera *Alexandrium* and *Gessnerium*. All toxin producing *Alexandrium* species belong to the subgenus *Alexandrium* with 14 listed species (Moestrup et al. , 2013); with *A. catenella, A. fundyense* and *A. tamarense* being responsible for the majority of PSP outbreaks on a global basis (Anderson et al., 2012). An interesting feature with *Alexandrium* is that several species consist of both toxic and non-toxic strains. Touzet et al. (2008) have demonstrated that there are both non-toxic and toxic strains of *A. minutum* have been found in Irish coastal waters. However, the non-toxic and the toxic forms do not co-exist, but are constrained to a western and a southern distribution, respectively (Touzet et al., 2008). According to Anderson et al. (2012), this is a general rule, but as they mention there are exceptions. Moreover, they refer to two known exceptions, from Northern Ireland and the Shetland Islands in Scotland where both toxic and non-toxic species in the *A. tamarense* complex have been documented.

Furthermore, four *Alexandrium* species have been demonstrated to produce other toxins than STXs, i.e. spiroloides and goniodomins, which are fast-acting macro-cyclic imines. Spirolides are produced by *Alexandrium ostenfeldii* and *A. peruvianum;* and reported to accumulate in shellfish from the North Atlantic, the Mediterranean and New Zealand (Suikkanen et al., 2013). Even if spirolides accumulate in shellfish there are no reports on human intoxications due to spirolides (Anderson et al., 2013). Moreover, Suikkanen et al. (2013) have, based on their recent work, suggested that wherever *A. ostenfeldii* and *A. peruvianum* will produce paralytic shellfish toxins, spirolides or both is a genetically fixed character linked to their phylogeny and biogeographic origin. *A. hiranoi* and *A. monilatum* are the producers of the latter toxin group, the gonidomins. They are ichthyotoxic and are reported to cause paralysis and mortality in finfish, but there are no reports on human intoxications (Anderson et al., 2012).

In Ireland, azaspiracids (AZAs) has become a major concern for the shellfish industry, with the first problems occurring in 1995 with human intoxications after consumption of toxic mussels produced in Irish waters (James et al., 2002). AZAs were also detected in shellfish from other parts of Europe; i.e. Norway, England, France and Spain (James et al., 2002; Magdalena et al., 2003). Furthermore, edible crab (*Cancer pagurus*) from Scandinavian waters has been shown to act as a vector for AZA (Torgersen et al. 2008). *Protoperidinium crassipes* was first pinpointed as the source organism (James et al. 2003), but recently the small dinoflagellate *Azadinium spinosum* was isolated during a research survey in the North Sea and shown to produce AZA (Krock et al., 2009; Tillmann et al., 2009). Further work has resulted in the identification of several species belonging to *Azadinium*, with reports from South America (Akselman and Negri, 2012). At present only *Azadinium spinosum* and *A. poporum* (Tillmann et al., 2011) have been demonstrated to produce AZAs. Recently, Jauffrais et al. (2012) were able to demonstrate through feeding experiments that blue mussels fed with toxic *A. spinosum* accumulated AZAs above the regulatory limit of 160 μg. kg^{-1}.

In Japan, in 1975 and 1980 food poisoning after consumption of the bivalva *Pinna pectinata* affected more than 1700 and 900 persons, respectively (see Uemura et al., 1995 and references therein). Uemura et al. (1995) and Chou et al. (1996) were the first to isolate and describe pinnatoxins, a new neurotoxin. Even though pinnatoxins have been known for several decades the causative species was unknown until recently. Rhodes et al. (2010) were able to demonstrate that a cyst-forming peridinoide dinoflagellate isolated from the waters of New Zealand is a producer of pinnatoxin. Nézan and Chomérat (2011), however, were the first to describe a novel dinoflagellate *Vulcanodinium rugosum* that later has been identified as the producer of pinnatoxins in Australia, New Zealand, Japan and China (Smith et al. 2011; Rhodes et al. 2011a; Rhodes et al., 2011b). Pinnatoxins have also been detected in shellfish from Norway (Rundberget et al., 2011), Canada (McCarron et al., 2012), and France (Hess et al., 2013). There are still uncertainties on pinnatoxins and acute toxicity towards humans (McNabb et al., 2012; Munday et al., 2012), hence further research is necessary to elucidate wherever pinnatoxins is a food safety issue and, whether there is a need for future regulations.

Harmful events caused by blooms of toxin producing naked or unarmoured dinoflagellates have been reported from all over the world; and we have chosen to focus on the three genera *Karenia, Karlodinium* and *Cochlodinum* recently reviewed in a special issue of *Harmful Algae.* Brand et al. (2012) have reviewed the present knowledge status on *Karenia* (previously *Gymnodinium*), which today consists of 12 recognized species. A wide variety of toxins with different toxic principles are demonstrated to be produced by the ten acknowledged harmful *Karenia* species (Moestrup et al., 2013), with brevetoxin, gymnocins, gymnodimine, and polyunsaturated fatty acids (PUFAs) among the described toxins so far (Brand et al., 2012). *K. brevis* produce brevetoxin that accumulates in fish and shellfish, and is the causative of neurotoxic shellfish poisoning (NSP) in Florida and Gulf of Mexico. The establishment of a monitoring program banning areas for harvesting of shellfish has been effective and human intoxications are now rare. Brand et al. (2012), however, report that the *Karenia* was transported from Florida to North Carolina with the Gulf Stream, resulting in an unexpected outbreak of NSP due to the lack of a monitoring program for *Karenia* in North Carolina. Furthermore, the first reported NSP outbreak outside U.S.A. occurred in New Zealand in 1993; *K. concordia* being the causative species. As with the case in North Carolina the bloom of *Karenia* in New Zealand was unexpected; and monitoring programs for the detection of *Karenia* and NSP were not established. Blooms of *K. brevis* can also be harmful in another way, since the cells are unarmoured they are more sensible towards turbidity, and can easily break releasing the toxins into the water or to the air as aerosols. Humans breathing brevetoxin-contaminated air experience respirational problems, which can be so severe that there is need for hospitalization (Brand et al., 2012). There are also reports on respirational problems due to *Karenia* species from other parts of the world, i.e. New Zealand and South Africa (see Brand et al., 2012)

Karenia blooms do not only affect human health, they may also have a severe impact on animal life with reported killings of both wild and cultivated fish, shellfish and other invertebrates, birds, turtles as well as marine mammals. In Europe, *Karenia mikimotoi* (previously described as *Gyrodinium aureolum*) was first recorded in Norway in 1966, and blooms have later on been reported from the waters off the U.K., Ireland and the Atlantic coast of France (Smayda, 2006). Dense blooms of *K. mikimotoi* along the Norwegian Skagerrak coast with subsequent fish mortalities was a recurring challenge for salmon farms in the 1970s and 1980s. Furthermore, blooms of *K. mikimotoi* have also killed farmed fish in Irish and Scottish waters.

Karlodinium (previously *Gymnodinium*) consists of eight species, with *K. armiger, K. corsicum* and *K. veneficum* listed as HAB species; i.e karlotoxin producers; by Moestrup et al. (2013). Place et al. (2012) reports on fish and shellfish mortality in connection with blooms of *K. veneficum* in Australia (fish) and Namibia (fish); and *K. corsicum* in Spain (cultivated fish and shellfish).

Cochlodinium is a large genus with approximately 40 accepted species of which only *C. polykrikoides* and *C. fulvescens* are reported to be toxic (Kudela and Gobler, 2012). The geographic distribution of *C. polykrikoides* was until recently restricted to U.S.A., China, Japan and Korea; but since the 1990s *Cochlodinium* has spread across Asia, North America, Australia and Europe (Kudela and Gobler, 2012 and references therein). *Cochlodinium* is ichythotoxic, and HABs of *Cochlodinium* have caused extensive fish mortalities in Japan and Korea (affecting local fisheries), as well as in the Canadian salmon industry and in Australia. Furthermore, Kudela and Gobler (2012) also refers to a bloom of *C. polykrikoides* in the Arabian Gulf and the Gulf of Oman in 2008 and 2009 that had severe effects not only by killing fish and coral reefs, but which also disturbed local desalination plants. Even if *C. polykrikoides* is recorded in the Mediterranean Sea, to the authors' knowledge, there have yet not been any reports on harmful events from European waters. Throughout the years there has been a lot of effort among researchers to identify the toxic principles and mechanisms of *Cochlodinium*. Several mechanisms, such as reactive oxygen species (ROS), neurotoxins and haemolytic toxins, have been suggested; but according to Kudela and Gobler (2012) the toxicity of *Cochlodinium* is yet not understood. *Cochlodinium* has, however, been demonstrated to degenerate fish gill lamella, and to pose negative effects on shellfish tissue.

Among the benthic dinoflagellates, there are several toxin producing species that pose a threat to human health either through the consumption of seafood or health effects due to toxic aerosols. Parsons et al. (2012) have recently reviewed the literature on the two genera *Gambierdiscus* and *Ostreopsis* focusing on taxonomy, geographical distribution, ecophysiology, and toxicology. *Gambierdiscus toxicus* was the first species to be identified as the producer of ciguatera toxin and maitotoxins, the toxins responsible for ciguatera food poisoning (CFP). *G. toxicus* and CFP are reported from the Pacific, the West Indian Sea and the Caribbean Sea; and CFP has for a long time been regarded as an endemic disease in these areas. Recent records of *Gambierdiscus* from the Canary Islands and the Mediterranean, as well as reports on CFP from the Canary Islands, Madeira and CFP-like episodes from the Mediterranean have, however, changed this perception and there is now an increasing concern on tropical and subtropical species spreading south- and northward from endemic areas (Parsons et al., 2012; Hallegraeff, 2010). Moreover, phylogenetic studies on *G. toxicus* isolated from different geographic areas have shown that there are at least fourteen species with different distribution; with at least three toxic and three non-toxic species, while the toxicity of the last eight species is yet to be elucidated (Parsons et al., 2012). From a European perspective, these findings have to be considered as a food safety issue; especially with respect to regulatory monitoring and risk assessment.

The *Ostreopsis* species concept has also been through a complete taxonomic revision, from the previously three described species to a number of nine recognized species (Parsons et al. 2012). Moreover, in addition to these nine species, there are still some records that are in need of further investigations to clarify wherever they are distinct species or not. As summarized by Parsons and co-workers (2012), *Osteropsis* is recorded in the Indian Ocean, Southeast Asia and Australasia, the Pacific and the Caribbean Sea, as well as in the Mediterranean Sea and the Eastern Atlantic region. The first record of *Ostreopsis* in the Mediterranean Sea was done in France in 1972, and later on *Ostreopsis* is reported from both European and North-African countries bordering the Mediterranean. Palytoxins, produced by species belonging to *Ostreopsis*, accumulate in seafood (i.e. crabs and fish). There are several reports on human seafood poisoning after consumption of contaminated seafood (Parsons et al. 2012 and references therein). A study from the Mediterranean is referred in the latter reference, that clearly indicates that a connection between high abundance of *Ostreopsis* and accumulation of palytoxins in shellfish. Aerosols produced by *Ostreopsis* have been set in connection with hospitalization of more than 150 people in Italy. However, whether the causative toxin is palytoxins or a new, undescribed toxin is yet not clear (Parsons et al., 2012). Furthermore, *Ostreopsis* have by some researchers been associated with ciguatera fish poisoning, but further studies are needed to confirm wherever there is such a connection.

Haptophytes

Haptophytes, or prymnesiophytes, are known from both marine and freshwater systems. According to Anderson (1992) there are about 500 recognized species of haptophyta, and an estimated number of species are 2 000. Haptophytes have two flagella and are named after the haptonema, an appendage used for feeding or attachment. The haptonema can be short and stiff, long and coiling, or rudimentary. All haptophytes have organic or calcareous scales, the latter belonging to the coccolithophorides, of which the scale structures have been important for the classification into species. The taxonomy of haptophytes was recently revised by Edvardsen et al. (2011) with introduction of new genera, and transferring of species between genera.

Blooms of haptophytes can be conspicuous due to their shear biomass, i.e. recognised by either discolouration (coccolithophorides) or foaming and a gelatinous appearance (*Phaeocystis*). The scales of coccolithophorides scatter light, and during bloom situations the sea is commonly reported as turquoise or of a milky-white appearance. Moreover, the coccolithophoride *Emiliania huxleyi* forms massive blooms in the northern Atlantic during summer, the geographic distribution is easily recognised on satellite images (Figure 4.1). Moreover, blooms of haptophytes have gained an interest from the research society due to the ability of *Emiliania huxleyi, Prymnesium parvum* and *Phaeocystis pouchetii* to

Figure 4.1. *Emiliania huxleyi* bloom off the Norwegian coast in 2006. Satellite image reproduced by permission of the European Space Agency, ESA (owner of copyright).

produce dimethylsulphoniopropiontes (DSMP) a biological precursor of dimethylsulphide (DMS). DMS from the oceans is the main natural source of atmospheric sulphur; and equals one-third of global anthropogenic sulphur emissions (Simo 2001). Atmospheric DMS contributes to Earth's albedo, either as aerosols or cloud condensations nuclei, and has been suggested as a link between global biosphere and the climate (Simo 2001 and references therein).

Species belonging to *Prymnesium* are associated with fish mortality all over the world, and the first recorded fish kill caused by *Prymesium* was reported in the Netherlands in 1920 (Moestrup, 1994). Moestrup also refers to episodes of fish mortality along the Baltic coast as far back as in the 1890s that later have been associated with *P. parvum.* He also mentions that fish kills caused by *Prymnesium* species have been reported from several

European countries, i.e. Denmark, United Kingdom, Finland, Bulgaria and Norway, during the previous century. In 2004, there was an extensive bloom of *Prymnesium* in Lake Koronia, Greece, with not only reports on fish mortality, but also bird kills (Michaloudi et al., 2009). According to the authors, this was the first recorded death of birds due to *P. parvum*. However, Lindholm et al. (1999) reported that dead gulls had been observed by locals during the *P. parvum* bloom in Lake Vargsundet, Finland, in 1997. Zingone et al. (2006) refer to an event with fish mortality due to *P. pateliferum,* now *P. parvum* f. *pateliferum* (Larsen 1999), from Italy. *P. parvum* blooms and fish kills have also been reported from Hungarian fish ponds (Vasas et al., 2012). Haemolytic toxins, prymnesins, are isolated from *P. parvum* and are identified as the toxic principle inducing fish mortality.

The bloom of *Prymnesium polylepis* (syn. *Chrysochromulina polylepis,* Edvardsen et al., 2011) in Kattegat and Skagerrak in 1988 has been thoroughly studied throughout the years, and in this paragraph we rely on the Smayda (2006) review of the extensive research material. The first indications on a HAB came from the Gullmar Fjord, Sweden, with reports on abnormal behaviour and mortality of farmed fish. When the bloom reached Norwegian waters, it killed 900 tonnes of farmed salmonides the negative effects of *P. polylepis* became evident. Towing of fish-farms from the affected coastal areas into fjords with lower salinity reduced the fish mortality, and thus turned out to be a successful mitigating activity. Another but not least important feature was the effect of the bloom on the ecosystem as a whole, with reports on mortality, not only in fish farms, but throughout the entire food chain including phytoplankton, zooplankton, benthic macroalgae and fauna, and demersal and pelagic fish. Professor Yasumoto and colleagues have demonstrated that *P. polylepis* produce two haemolytic and ichthytoxic compounds (Moestrup and Thomsen, 2003). The wide-range consequences of this bloom were in many ways a game changer increasing the awareness of HABs, as well as opening up for the establishment of regular algal monitoring programs.

A few years later, in May in 1991, a bloom of *Chrysochromulina leadbeateri* caused mortality of farmed fish in the Vestfjorden-Lofoten area in Northern Norway (Smayda, 2006). This was the first recorded bloom of this species. As for *Prymnesium parvum* and *Prymnesium polylepis* there is evidence that the toxicity of *C. leadbeateri* is due to the production of haemolytic compounds. In May 2008, unusual fish mortality was once again reported from the Lofoten area, and *C.* cf. *leadbeateri* was identified in samples collected from the affected localities (Arff and Tangen, 2009). Furthermore, Smayda also refers to mortality of rainbow trout in connection with a bloom of *Chrysochromulina brevifilum, C. ericina, C. hirta,* and *C. spinifera* in Danish waters in 1992. However, there was no clear evidence on the harmful principles of these four *Chrysochromulina* species none being demonstrated as toxin producers. *C. leadbeateri* is also reported from the Mediterranean (Zingone et al., 2006).

The colonial stage of *Phaeocystis globosa* and *Phaeocystis pouchetii* are reported to form nuisance blooms in European waters. Blooms of *P. globosa* are a well-known phenomenon off the coasts of the Netherlands and Germany, where characteristic foam produced by the algae assembles on beaches. In Germany, these algae are called "schaumalge" literally meaning "foam algae". *P. pouchetii* is common during the spring bloom of diatoms in Norwegian waters (www. Algeinfo.no), and especially in Mid- and

Northern Norway. Moestrup (1994) refers to episodes with fish kills in Norwegian waters in connection with toxic blooms of *P. pouchetii,* but at that time *Phaeocystis* was only known to produce antibacterial acrylic acid. Tangen (1999), however, reports that the fish killings caused by *P. pouchetii* in Norway not are considered to be caused by toxins, but rather was an effect of the mucilage irritating and clogging the gills. More recent both *P. globosa* and *P. pouchetii* are reported to produce compounds with haemolytic activity, but the ichtyhotoxic mechanisms differ with *P. globosa* being a glycolipid producer, while there still are some uncertainties with respect to *P. pouchetii* (Rijssel et al., 2007).

Raphidophyceae

Raphidophyceans are registered in both marine and freshwater systems, and their fish-killing potential is known from mariculture in Japan, New Zealand, Chile and Canada (Hallegraeff and Hara, 2003). *Heterosigma akashiwo, Chattonella* spp., and *Fibrocapsa japonica* were reported as causative species. Recently, a new *Raphidophyceae, Viridilobus marinus* (Demir-Hilton et al., 2012) was co-occurring with *Heterosigma* in coastal waters off U.S.A. Raphidophyceans are susceptible toward common fixatives as formaldehyde and Lugol's solution and the presence of these algae in preserved material are easily overlooked. Hence, their occurrence and geographical distribution may be unreported. Several toxic mechanisms have been suggested: reactive oxygen species (ROS), brevetoxin-like compounds, haemagglutination and haemolytic activity, as well as mucus excretion clogging the gills. Even if the ichthyotoxicity of raphidophytes still is a hot topic among researchers (Dorantes-Aranda et al., 2013; Kwak et al., 2012; Boer et al., 2012; Pezzolesi et al., 2010; Boer et al., 2009). The research community has yet to conclude on the fish-killing properties of Raphidophyceans.

Fish mortality and economical loss due to *Heterosigma akashiwo* are known from a wide range of countries with aquaculture industry. Some examples are the recurrent blooms in Japan in 1972-1987 and the bloom in Canada in 1997 extending over a period of four months (Smayda, 2006). The first reported mass occurrence of *H. akashiwo* in European waters is the bloom in the Oslofjord, Norway, in 1964. Later on several blooms of *H. akashiwo* with subsequent mortality in fish farms have been recorded from Spain, Norway, Ireland, and Scotland, as well as a report on bivalve mortality from Portugal (Smayda, 2006). Furthermore, there have been discussions on wherever *H. akashiwo* is an introduced or indigenous species in European waters. Based on available research, Smayda (2006) concludes that *H. akashiwo* belongs to the latter. A special feature of *H. akashiwo* is that it has a very wide niche. Hence, blooms have been recorded during different nutritional conditions, varying stratification, in areas with freshwater influence or high salinity, as well as in pristine areas and in aquaculture-rich areas (Smayda, 2006).

Recent phylogentic studies of *Chattonella* spp. have led to taxonomic revisions of this genera with *C. verruculosa* being transferred to the Dictyochophyceae (Hosoi-Tanabe et al. 2007; Edvardsen et al., 2007; Eikrem et al., 2009; Chang et al., 2012), and evidence that *C. ovata* and *C. antiqua* are varieties of *C. marina* (Demura et al., 2007). According to these findings *Chattonella* at present consists of three species; i.e. *C. subsalsa, C. minima,* and *C. marina* (with three varieties *marina, antiqua* and *ovata*). The first recorded HAB of *Chattonella* with fish mortality is from India; thereafter *Chattonella* has become infamous for massive killings of farmed fish in Indian, Japanese, Korean, and Australian waters, as well as in the waters off U.S.A. (Imai and Yamaguchi, 2012). Moreover, Imai and Yamaguchi refer to episodes where *Chattonella* has damaged local fisheries. Zingone et al. (2006) reports that *C. subsalsa* has been recorded in the Tyrrhenian Sea, but to their knowledge there have been no harmful events in that area.

Blooms of *Fibrocapsa japonica* are common in Japan with subsequent fish mortalities. The first evidence of *F. japonica* in European waters came from France and the Netherlands in 1991 and from Germany in 1993 (Smayda 2006). *F. japonica* is also reported in the Adriatic Sea and Tyrrhenian Sea (Cucchiari et al., 2010). So far there are no reports on *F. japonica* and fish-kills in European waters. Boer et al. (2009) have studied the possibility of future harmful events on fish in Dutch waters, and they conclude that climate change will contribute to an increased risk for development of harmful algal blooms. From the Adriatic Sea, however, discolouration from *F. japonica* blooms is reported to be nuisance towards tourism and recreational activities (Cucchiari et al., 2010).

Dictyochophyceae

In 1984, salmon and trout farmed in Danish waters died off during a bloom of the naked form of the silicoflagellate *Dictyocha speculum* (syn. *Distephanus speculum*); being the first report of *D. speculum* as a HAB species (see Smayda 2006 and references therein). Smayda also refers to mortality in mariculture in connection with blooms of siliceous *D. speculum* in Spain (salmon, turbot, and octopus), France (salmon) and Shetland (salmon). Moreover, he discusses wherever the bloom of the unidentified Flagellate X reported to be the cause of fish kills in Scottish lochs in 1979 and 1982 was a mixture of three known icththyotoxic flagellates; the naked *D. speculum, Heterosigma akashiwo* and *Chattonella* cf. *verruculosa,* the latter is as described above transferred to the Dictyochophycea and renamed *Pseudochattonella.*

The mechanism behind the fish-kills is not known, but both the siliceous and the naked form are believed to irritate the gills causing mucus production, oedema and hyperplasia, as well as hypoxia and hypercapnia. Smayda also refers to a HAB event in the Adriatic Sea with benthic mortality as a consequence of the decomposition of an extensive bloom of *D. speculum* creating anoxic conditions in the bottom waters. *D. speculum* has a complicated life cycle alternating between two motile stages, including a siliceous and a naked form, and a non-motile multinucleate stage. Smayda reports that the siliceous stage is more common in open, coastal waters, while the naked form tends to dominate in more enclosed, onshore areas. Interestingly, he has reviewed several papers on silicoflagellate blooms in European waters and the findings are all pointing to the same direction with mass occurrences developing in connection with nutrient enrichment of either terrestrial or anthropogenic origin. These studies have all demonstrated that nitrogen is the most important prerequisite for bloom development. Another interesting feature is that the alternating between the two stages makes *D. speculum* less dependent on silica than diatoms, hence *D. speculum* can bloom after the diatoms have become Si-depleted.

The first reported bloom of *Pseudochattonella* (Hosoi-Tanabe et al., 2007; Edvardsen et al., 2007; Eikrem et al., 2009), previously identified as the raphiodphyte *Chattonella* aff.

verruculosa, in European waters was in April-May in 1998 (Naustvoll, 2006). Naustvoll, however, reports that reanalysis of material from the Swedish west coast have shown that *Pseudochattonella*-like cells have been present in low concentrations in this area since 1993. Pettersson and Pozdnyakov (2013) have recently published an analysis of the blooms of *Pseudochattonella* in Skagerrak, Kattegat and the German bight during the last two decades, giving an overview of the blooming events from 1998 to 2009. In 1998, the bloom originated off the Danish west coast in April; and spread northwards along the Danish coast to Kattegat and Skagerrak, most of the bloom is, however, thought to have been advected directly from Denmark to southern parts of Norway. In May of the same year, Norwegian salmon-farmers located on the Skagerrak coast reported on increased fish mortality, and high concentrations of *Pseudochattonella* was found in samples collected in the proximity of the farming sites. Furthermore, there were reports on mortality of wild fish in Danish waters, including herring, mackerel and eels. Also in 2001, *Pseudochattonella* was blooming in Skagerrak/Kattegat with subsequent mortalities of farmed salmon in Norwegian coastal waters. Recently the recurrent blooms of *Pseudochattonella* have become a problem for the Danish aquaculture industry. As a consequence a monitoring program on the presence of *Pseudochattonella* (and other HAB species) in Danish waters has been established for risk assessment (Andersen, 2007).

Pseudochattonella in European waters is thought to be an alien species which is now established with recurrent blooms in both Skagerrak and Kattegat (Naustvoll, 2006). The work performed by Riisberg and Edvardsen (2009) have demonstrated that there were two distinct species during the bloom in 2000 with *P. farcimen* in Scandinavian waters and *P. verruculosa* in the German bight. Moreover, *P. farcimen* have so far only been identified in samples from Scandinavia, and Riisberg and Edvardsen indicates that *P. farcimen* might be indigenous; while their research supports that the European *P. verruculosa* is an introduced species. However, the need for further research on *Pseudochattonella* from different geographic regions is critical before concluding on wherever *P. farcimen* is an indigenous species in Skagerrak and Kattegat or not.

There are still uncertainties on the ichthyotoxic principles with which *Pseudochattonella* kill fish. A study by the Norwegian School of Veterinary Medicine on caged salmon exposed to *P. farcimen* during the bloom in 2001 set gill necrosis and bleeding in the orbital cavity in connection with the exposure to *Pseudochattonella* (Skjelbred et al., 2011). Skjelbred et al. (2011) investigated the ichthyotoxic mechanisms of *Pseudochattonella* spp. and concluded that there are indications on metabolites affecting fish cell lines and gills of cod and salmon larvae. Moreover, Dittami et al. (2012) have recently identified metabolic pathways that might be involved in the production of toxic compounds in *P. farcimen.*

Recently, *Chattonella globosa* (Raphidophyceae) was transferred to Dictyochophyceae and renamed *Vicicitus globosus* (Chang et al., 2012). Interestingly, Lømsland et al. (2013) reported on the co-occurrence of *Chattonella globosa, Dictyocha fibula* and algal cells having features indicating intermediate stages between these two species in connection with mortality in salmon-farms located in southern Norway in 2008. Furthermore, they suggest that *C. globosa* might be the naked form of *D. fibula.* Unfortunately, their observations were only documented by light microscopy. Imai and Yamaguchi (2012), however, refer to a Japanese study which seems to be the first to document the phylogenetic connection between *C. globosa* and *D. fibula*. Furthermore, according to the review on *Chattonella* by Imai and Yamaguchi (2012), these researchers suggested a transfer of *C. globosa* to the Dictyochophycea.

Cyanobacteria
There are freshwater, brackish and marine species among the cyanobacteria and the prevalence of cyanobacterial blooms are generally larger in freshwater bodies than in brackish and marine waters. Several cyanobacteria are toxin producers. The toxins can be grouped according to their target organ in mammals, i.e. hepatotoxins (microcystins, nodularins and cylindrospermopsis), neurotoxins (anatoxins, saxitoxins, palytoxins and beta-methylamino-L-alanine [BMAA]), skin (aplysiatoxins, lyngbyatoxin), gastro-intestinal tract (lyngbyatoxin), and irritants (lipopolysaccharides) (O`Neil et al., 2012). In the Baltic Sea and adjacent waters (i.e. Gulf of Finland and the Bothnia Sea), blooms of cyanobacteria are a common summer phenomenon. Bordering countries Denmark, Sweden and Finland have all established regular monitoring programs to follow the development and progress of cyanobacterial HAB events. The primary bloom-forming genera are the non-toxic *Aphanizomenon* and the toxin-producers *Nodularia* and *Anabena*; with *N. spumigena* being the dominating species. There are several reports on animal intoxications from the Baltic Sea. A study on the possibility of fish acting as a vector for toxin transfer through the food web Sipiä et al. (2007) found indications that three-spined stickleback (*Gasterosteus aculeatus*) could act as a vector.

Strictly marine cyanobacteria are found among the genera *Lyngbya*, *Trichodesmium* and *Synechococcus*. However, blooms of marine cyanobacterial species are only reported from tropical and sub-tropical waters. Generally *Lyngbya* spp. are benthic mat forming species, while species belonging to *Tricodesmium* and *Synechococcus* are pelagic. *L. majuscula* is reported to form HABs; with the first incident in Hawaii in the 1950s, and more recent in waters off Australia, Florida and the Caribbean (O´Neil et al., 2012). As reviewed by O´Neil and colleagues (2012), humans exposed to HABs of *L. majuscula* report on dermatitis and asthma-like symptoms. Furthermore, *L. majuscula* blooms may have severe consequences; Yasumoto (1998) demonstrated that consumption of poisonous green sea turtle meat (lyngbyatoxin A) was the cause of a fatal intoxication in Madagascar. *Synechococcus* is cosmopolitan, with long-lasting blooms that can cover large areas (O´Neil et al., 2012). Its blooms may have a negative impact on the ecosystem as a whole (O´Neil et al., 2012). Anoxic conditions, increased light attenuation impacting benthic communities negatively, and inhibition of zooplankton grazing are reported. Previously only freshwater species of *Synechococcus* were known to be toxin producers, but it was found that that a haline strain of *Synechococcus* is able to produce microcystin (Carmicheal and Li, 2006).

Trichodesmium may form large blooms in oligotrophic waters, with reports on blooms from the Pacific, Arabian Sea, Indian Ocean, Gulf of Mexico and the Caribbean (O´Neil at al. 2010). An Australian study (Negri et al., 2004) implicates that the observed mortality among farmed pearl oysters was caused by malnutrition during a large *Trichodesmium* bloom, and not as first suspected from toxins produced by the cyanobacteria. There are also reports that anoxic conditions due to decaying *Trichodesmium* have resulted in mortality among marine organisms (i.e. fish, shrimp and oysters) (Negri et al., 2004). There are reports on symptoms to humans after exposure to *Trichodesmium* with headaches, skin-irritations and asthma. Moreover, ciguatera-like intoxications have been reported from Australia after consumption of mackerel (O´Neil et al., 2012). Recently, palytoxin and a derivate have been isolated from both *T. erythraeum* and *T. thiebautii* (O ´Neil et al., 2012). This finding suggest that not only dinoflagellates, but also cyanobacteria might be involved in CFP (O´Neil et al., 2012). There are also indications that *Trichodesmium* produce other toxins; i.e. microcystins, cylindrospermopsin and saxitoxin (O´Neil et al., 2012).

Oceanographic conditions and HABs

HABs are very complex phenomena, present virtually in every coastal country in the world and they can occur under a great range of oceanographic conditions. This complexity, together with the non-straightforward interaction between physical oceanographers and biologists made it difficult to establish a clear link between the appearance of blooms and those conditions. The first studies on the subject were mostly focused on algal taxonomy, toxin chemistry and phytoplankton ecology (Anderson et al., 2010). After a number of years of research, it became eventually clear that a different approach was needed. This new approach required carrying out international, multidisciplinary and comparative studies as suggested in the Global Ecology and Oceanography of Harmful Algal Blooms Program (GEOHAB) implementation plan (GEOHAB, 2003). The activities carried out in the framework of this program have been instrumental for these purposes.

GEOHAB is an international program of the Scientific Committee on Oceanic Research (SCOR) and the Intergovernmental Oceanographic Commission (IOC) of the United Nations Educational, Scientific and Cultural Organization (UNESCO) The GEOHAB program was developed around four ecosystem types which were defined on the basis of their oceanographic regime: (1) fjords and coastal embayments, (2) upwelling systems, (3) eutrophic systems and (4) stratified systems. This approach has proven helpful to improve our understanding of the interactions between physical and other environmental conditions leading to the occurrence of blooms. Hence, in the following section we will be using that classification to describe the state-of-art concerning the relationships between oceanographic conditions and HABs, providing European examples to illustrate each ecosystem type.

Fjords and coastal embayments

Without being exhaustive, these aquatic systems can be defined as enclosed or semi-enclosed environments along a land-mass margin (Cembella et al., 2005). Their hydrographic regimes are heterogeneous and influenced by many factors such as estuarine circulation, the presence of fronts and internal waves or the coupling to offshore circulation. Despite their heterogeneity, these systems can be characterized by a number of common features that derive, precisely, from their confinement and the importance of land-sea interactions. GEOHAB (2010) relates four as the most standing ones: (1) the importance of residence time, (2) the exchange with off-shore water, (3) the influence of freshwater discharge and (4) the increased pelagic-benthic coupling. Getting insight into these features is needed for the understanding of HAB dynamics in such confined environments. One very important question has to do with exogenous vs. endogenous origin of the bloom. One illustrative example of the first case takes place in Flødevigen bay, along the southern coast of Norway, where phytoplankton assemblages can be advected by onshore currents (more in particular, the Norwegian Coastal Current) and then trapped in the bay generating HABs (GEOHAB, 2010). Another good example is Bantry Bay, in Southern Ireland, where the appearance of *Dinophysis acuminata* is determined by the wind-driven exchange of water within the bay, a two-layer oscillatory flow acting in a thermally stratified water column and bringing the toxic phytoplankton inside it (Raine et al., 2010). For the endogenous case, i.e. if the embayment acts a seed bed, the blooms will be more cyclic (seasonal) and less dependent on circulation patterns and hydrographic variability along the outer coast (Cembella et al., 2005).

Upwelling systems

Upwelling areas are characterized by an influx of deep nutrient-rich waters to the photic zone. This influx is originated by the wind blowing equatorward along the eastern boundaries of the oceans. The Ekman transport associated with this wind transports the superficial layer offshore and this layer is replenished by the upwelling of colder, nutrient-rich waters near the coast. Consequently, these areas can present a high productivity and be prone to the development of HABs. The timing of those HABs in upwelling systems is determined by wind stress changes and freshwater inputs at seasonal events and interannual scales (Pitcher et al., 2010). In turn, their location is mostly controlled by inner-shelf dynamics related to changes in coastline morphology and bathymetry. The development of HABs within upwelling systems is particularly favored in open embayments because they provide enough nutrients while remaining stratified and accessible to the introduction, concentration and retention of coastal blooms.

The Rías Baixas, located on the NW coast of Spain and part of the Iberian upwelling system, is a good example of this type of embayments. They are four large and narrow inlets whose high-productivity makes them an exceptional site for the culture of mussels, hence their economic importance. The type of harmful species appearing changes along the upwelling cycle, which is seasonal and determined by the latitudinal shift of the trade wind system (GEOHAB, 2005). Blooms of diatoms such as the genus *Pseudo-nitzschia* that can produce ASP toxins are mostly linked to pulses of upwelling which are more frequent in summer period. Autotrophic dinoflagellates like *Gymnodinium catenatum* and *Alexandrium minutum* are more abundant in late summer when the upwelling relaxes, following the increase of stratification and taking advantage of their swimming capabilities (Trainer et al. 2010). A relationship between the renewal time of water within the Rías and the number of days, at which mussel extraction is interrupted due to the presence of harmful microalgae, has been suggested (Alvarez-Salgado et al., 2008). This is more pronounced for dinoflagellates of the genus *Dinophysis*. During upwelling events, the positive residual circulation within the Rías is intensified, transporting dinoflagellates to the shelf. When upwelling relaxes, the circulation of water within the Rías slows down (larger renewal time), dinoflagellates are advected to the inner part of the Rías and HABs can occur, with the subsequent negative impact on local economy.

Eutrophic systems

The high nutrient concentration characteristic of eutrophic systems explains why they are prone to HABs formation. In many cases, this has to do with a more intense nutrient loading of human origin, which is thought to be one of the reason for the intensification of HABs episodes in the last decades (GEOHAB, 2006). Nevertheless, the mechanisms underlying this relationship are not totally understood and go far beyond the simplistic notion of "more nutrients higher algal biomass". Research over the last decades has enabled to make considerable progress in this field, as shown in the special issue of *Harmful Algae* (Glibert et al., 2008). In this issue, Heisler et al. (2008) presents some consensus statements including (1) the proven relationship between degraded water quality and the persistence of HABs and (2) the impact of nutrient pool composition (not just the total quantity). For the latter there are examples of HAB species responding to pulses of nutrients but also to a more sustained, chronic input.

From a European perspective, a good example is the northwestern Black Sea. This area experienced an intense nutrient loading in the 1970s and 1980s, with an increase to the number of toxic blooms, replacing of diatoms and larger algae with dinoflagellates and nanoplankton (Anderson, 2002). On the other hand, the European Marine Strategy Framework Directive (MSFD, 2008/56/EC) includes eutrophication as one of the qualitative descriptors for determining good environmental status, mentioning harmful algae blooms amongst its adverse effects. The work of the eutrophication task force created by the European Commission to develop guidance in this field has been reviewed (Ferreira et al., 2011), with several examples of blooms related to eutrophic conditions and waterborne nutrient inputs in Europe.

Stratified systems

Stratified waters can be found in all afore-mentioned systems (upwelling, coastal, *etc*) and in this sense they are applicable to them all. An article of Berdalet et al. (2013a) provides a review in the progress in this field. In stratified waters, populations of HABs can appear and concentrate in thin (i.e. from decimeters to a few meters thick) and subsurface layers, extending horizontally up to several kilometers (Sullivan et al., 2010). These can persist from hours to weeks. The outbreak of blooms in these environments can be favored through several mechanisms. For instance, these thin layers can appear in the upper part of the nutricline, thus providing photosynthetic harmful algae with nutrients and sufficient light levels. Furthermore, even if the conditions are not particularly good for individual cell growth, total growth can increase because the dispersive losses are limited in thin layers. This relatively high concentration of cells can produce chemical substances that hinder grazing. Finally, growth rates can also be enhanced by a greater probability of sexual encounters (GEOHAB, 2008).

In Europe, it is possible to find several examples of this type of interactions and the different mechanisms that originate such thin layers. In the Rías Baixas (NW coast of Spain), thin layers with populations of *Pseudo-nitzschia* spp. and *Chaetoceros socialis* were observed (Velo Suarez et al., 2010). Differences in shear profiles during an upwelling event led to the formation of the thin layers where the toxin-producing diatoms developed. Raine et al. (2013) review the effect of physical factors such as tides, solar heating and winds on the promotion of HABs around Ireland, along with a comprehensive description of their consequences over a range of length scales from hundreds of kilometers to a few meters.

Monitoring of algal toxins in Europe

In Europe, monitoring of algal toxins in shellfish is regulated by EC directives and guidelines. Regulations set maximum levels for paralytic shellfish poisoning (PSP) toxins (e.g. saxitoxin), diarrhetic shellfish poisoning (DSP) lipophilic toxins (including the okadaic acid group and the pectenotoxin group), ASP toxins (e.g. domoic acid) as well as two other groups of lipophilic marine toxins: the azaspiracid group (AZA) and the yessotoxin group (YTX). Regulation 854/2004/EC mandates the competent authorities in the EU Member States to examine live bivalve molluscs for the presence of marine biotoxins before being placed on the market. With regard to methodologies, Annex III of EC Regulation No. 2074/2005/EC established recognised testing methods for marine biotoxins, including biological methods (mouse bioassay and rat bioassay) in the case of PSP and lipophilic toxins, as reference methods and chemical methods (for ASP toxins). This has recently been amended by EC Regulation No. 15/2011. This new regulation establishes that the reference method for the detection of lipophilic toxins should be a validated technique of liquid chromatography and mass spectrometry, available for consultation on European Union Reference Laboratory, EU-RL. The EU-RL ensures a standardized system throughout the EU by coordinating the European Network of National Reference Laboratories. In addition to monitoring of biotoxins, most European countries also have a regular monitoring program for the detection of the corresponding source organisms. Furthermore, research has demonstrated that the toxin content of source species depend on several factors and among them geographical differences have been identified (Anderson et al., 2012). Therefore regulatory limits for the toxin producing microalgae may differ between European countries.

Parades et al. (2011) have recently reviewed the current European monitoring program on marine biotoxins, and suggest a revision of the marine biotoxins regulated in the EU legislation. Their concerns are especially directed to non-regulated toxins, i.e. tetrodotoxin, palytoxin and some cyclic imines; that have been detected in European shellfish. Furthermore, they discuss the need to deregulate toxins that have been shown not to pose a risk for EU's population (i.e yessotoxin).

Conclusions and discussion

During the past two decades monitoring and research programs have demonstrated that HAB species are spreading to new areas. This is also evident in European waters. Toxin-producing species that previously were reported to have a tropical or sub-tropical distribution are now recorded in the Mediterranean Sea. Several of these biotoxins are not included in European monitoring programs; thus composing a threat towards humans consuming seafood. Furthermore, we foresee the need for new or improved detection methods that not only give information on wherever toxins are present or not, but also are able to identify the chemical structure of novel or emerging toxins.

Future trends

From a global perspective, a growing population puts on the agenda the need for a stable supply of healthy and nutritious food. The Food and Agriculture Organization of the United

Nations (FAO) has, in this context, stated that the fisheries and aquaculture sector offers opportunities to increase food and nutrition security, alleviate poverty, generate economic growth and ensure improved use of resources. Stimulated by the higher demand for seafood they have projected that the world fisheries and aquaculture production will reach about 172 million tons per annum in 2021, with most of the growth in the aquaculture sector.

However, there is an increasing concern regarding climate change and how it will affect phytoplankton, HAB species especially and the implications this might have to human health (Balbus et al., 2013; Hallegraeff, 2010; Hays et al., 2005). Climate change may impact to the geographical distribution of phytoplankton, timing of blooms, as well as have secondary effects to phytoplankton grazers. Furthermore, Hinder et al. (2011) have performed an analysis of the effects climate change may have on HABs and shellfish safety in the U.K. They concluded that a potential future climate change could bring new toxic species into the waters surrounding the U.K. Thus, climate change might lead to changes in species composition, as well as to the introduction of new and potentially toxic species that the present detection systems cannot monitor.

To meet the forthcoming demands for an expanding aquaculture sector producing safe seafood to a growing population in an era with climate change, which is foreseen to affect the spatial and temporal distribution of HABs, the complexity of monitoring programs will probably increase. Further research and knowledge on marine biotoxins, their producers, and under which oceanographic conditions they become harmful will be essential to meet these requirements. Furthermore, the development of HAB species specific sensors, as well as biotoxin specific sensors that can be deployed on near real-time monitoring systems will play an important role in future HAB monitoring and seafood safety. Especially, if deployed in areas with seafood production near real-time monitoring systems can provide early warnings on the development of HAB events to regulatory authorities and the producers. This will contribute to a better risk assessment and more precise regulation of the resources, as well as to the implementation of mitigation strategies if existing. Moreover, HAB sensors deployed on near real-time monitoring systems will, together with standard oceanographic sensors, give valuable input to oceanographic models and future development of reliable HAB forecasts based on such measurements.

Web resources

- IOC-UNESCO taxonomic reference list of harmful microalgae; http://www.marinespecies.org/HAB
- AlgaeBase; http://www.algaebase.org
- Nordic microalgae and aquatic protozoa; http://nordicmicroalgae.org/
- The European Union Reference Laboratorty for Marine Biotoxins; http://www.aesan.msssi.gob.es/en/CRLMB/web/home.shtml
- Global Ecology and Oceanography of Harmful Algal Blooms; http://www.geohab.info/
- The Intergovernmental Oceanographic Commission of UNESCO Harmful Algal Bloom Program; http://ioc-unesco.org/hab/
- Information on the algal situation in the Baltic Sea; http://www.smhi.se/en/Weather/Sweden-weather/the-algae-situation-1.11631; http://www.dmi.dk/en/hav/satellitmaalinger/algal-maps/

http://www.itameriportaali.fi/en/tietoa/algaline_seuranta/en_GB/algaline_seuranta/
• Information on the algal situation in Irish coastal waters; http://www.marine.ie/home/publicationsdata/data/Habs+Search+Database/PhytoplanktonShellfishToxicitySummary.html
• Information on the algal situation in Norwegian coastal waters; http://algeinfo.imr.no/
• Information on shellfish safety in Norway; http://www.matportalen.no/verktoy/blaskjellvarsel/

References

Akselman, R., Negri, R.M. (2012). Blooms of *Azadinium* cf. *spinosum* Elbrächter et Tillmann (Dinophyceae) in northern shelf waters of Argentina, Southwestern Atlantic. Harmful Algae *19*, 30–38.

Álvarez–Salgado, X.A., Labarta, U., Fernández–Reiriz, M.J., Figueiras, F.G., Rosón, G., Piedracoba, S., Filgueira, R., Cabanas, J.M. (2008). Renewal time and the impact of harmful algal blooms on the extensive mussel raft culture of the Iberian coastal upwelling system (SW Europe). Harmful Algae *7*, 849–855.

Andersen, P. (2008) Videnbaseret operationel overvågning af skadelige alger i forbindelse med danske havbrug. Status for forekomsten af skadelige alger i foråret 2006. Enviromental Report. Orbicon Consultancy, 1-38. (In Danish).

Anderson, D.M, Glibert, P.M., Burkholder, J.M. (2002). Harmful algal blooms and eutrophication: nutrient sources, composition and consequences. Estuaries *25*, 702–726.

Anderson, D.M., Reguera, B., Pitcher, G.C., Enevoldsen, H.O. (2010). The IOC International Harmful Algal Bloom Program: History and science impacts. Oceanography *23(3)*, 72–85.

Anderson, D.M, Alperman, T.J., Cembella, A.D., Collos, Y., Masseret, E., Montersor, M. (2012). The globally distributed genus *Alexandrium*: Multifaceted roles in marine ecosystems and impacts on human health. Harmful Algae *14*, 10-35.

Andersen, R.A. (1992). Diversity of eukaryotic algae. Biodivers. Conserv. *1*, 267-292

Arff, J., Tangen, K. (2009). *Chrysochromulina* cf. *leadbeateri* in Northern Norway. Poster presented at GEOHAB Modeling Workshop, 15-19 June, Galway, Ireland, Marine Geological and Biological Habitat Mapping.

Balbus, J.M., Boxall, A.B.A., Fenske, R.A., McKone, T.E., zeise, L. (2013). Implications of global climate change for the assessment and managment of human health risks of chemicals in the natural environment. Environ. Toxicol. Chem. *32*, 62-78.

Bates, S.S., Bird, C.J., de Freitas, A.S.W., Roxall, R., Gilgan, M., Hanic, L.A ., Johnson, G.R., McCulloch, A.W., Odense, P., Pocklington, R., Quilliam, M.A, Sim, P.G., Smith, J.C., Subba Rao, D.V., Todd, E.C.D, Walter, A.J., Wright, J.L.C (1989). Pennate diatom *Nitzschia pungens* as the primary source of domoic acid, a toxin shellfish from Eastern Prince Edward Island, Canada. Can. J. Fish. Aquat. Sci. *46(7),* 1203-1215.

Berdalet, E., McManus, M.A., Ross, O.N., Burchard, H.,Chavez, F.P., Jaffe, J.S., Jenkinson, I.R., Kudela, R., Lips, I., Lips, U., Lucas, A., Rivas, D., Ruiz-de la Torre, M.C., Ryan, J., Sullivan, J.M., Yamazaki, H. (2014). Understanding harmful algae in stratified systems: review of progress and identification of gaps in knowledge. Deep-Sea Res. Pt. II *101*, 4-20.

Boalch, G.T., (1987) Changes in the phytoplankton of the Western English Channel in recent years. Br. Phycol. J. *22(3)*, 225-235

Boer, M.K.d. Tyl, M.R., Meng Fu, M., Kulk, G., Liebezeit, G., Tomas, C.R., Lenzi , A., Naar, J., Vrieling, E.G., Rijssel, M.v. (2009) Haemolytic activity within the species *Fibrocapsa japonica* (Raphidophyceae). Harmful Algae *8*, 699–705

Boer, M.K.d Boerée, C., Sjollema, S.B., de Vries, T., Rijnsdorp, A.D., Buma, A.G.J. (2012) The toxic effect of the marine raphidophyte *Fibrocapsa japonica* on larvae of the common flatfish sole (*Solea solea*). Harmful Algae *17*, 92–101

Bogan, Y.M., Kennedy, D.J., Harking, A.L., Gillespie, J., Vause, B.J., Beukers-Stewart, B.D., Hess, P., Slater, J.W., (2007a). Variation in domoic acid concentration in king scallop (*Pecten maximus*) from fishing grounds around the Isle of Man. Harmful Algae *6*, 81-92.

Bogan, Y.M., Bender, K., Hervas, A., Kennedy, D.J., Slater, J.W., Hess, P. (2007b). Spatial variability of domoic acid concentration in king scallops *Pecten maximus* off the southeast coast of Ireland. Harmful Algae *6*, 1-14.

Brand, L.E., Campbell, L., Bresnan, E. (2012) Karenia: The biology and ecology of a toxic genus. Harmful Algae *14*, 156–178

Carmichael, W.W., Li, R.(2006) Cyanobacteria toxins in the Salton Sea. Saline Systems *2(5)*, 1746-1448.

Cembella, A.D., Ibarra, D.A., Diogene, J., Dahl, E. (2005). Harmful algal blooms and their assessment in fjords and coastal embayments. Oceanography *18(2)*, 158–171.

Cembella, A., Guzmán, L., Roy, S., Diogène, J. , eds. (2010). Global ecology and oceanography of harmful algal blooms: HABs in fjords and coastal embayments. (Paris, France and Baltimore,U.S.A.: Intergovernmental Oceanographic Commission-UNESCO and Scientific Committee on Oceanic Research).

Chang, F.H., McVeagh, M., Gall, M., Smith, P. (2012). *Chattonella globosa* is a member of Dictyochophyceae: reassignment to *Vicicitus* gen. nov., based on molecular phylogeny, pigment composition, morphology and life history. Phycologia *51(4)*, 403-420.

Cucchiari, E., Pistocchi, R., Pezzolesi, L., Penna, A., Battocchi, C., Cerino, F., Totti, C. (2010). Resting cysts of *Fibrocapsa japonica* (Raphidophyceae) from coastal sediments of the northern Adriatic Sea (Mediterranean Sea). Harmful Algae *10*, 81–87.

Demir-Hilton, E., Hutchins, D.A., Czymmek, K.J., Coyne, K.J. (2012). Description of *Viridilobus marinus* (gen. et sp. nov.), a new raphidophyte from Delaware's inland bays. J. Phycol. *48*, 1220–1231.

Demura, M., Noel, M.H., Kasai, F., Watanabe, M.M. Kawachi, M. (2009).Taxonomic revision of *Chattonella antiqua, C. marina* and *C. ovata* (Raphidophyceae) based on their morphological characteristics and genetic diveristy. Phycologia *48(6)*, 518-535.

Dittami, S.M, Riisberg, I., John, U., Orr, R.J.S, Jakobsen, K.S., Edvardsen, B. (2012) Analysis of expressed sequence tags from the marine microalga *Pseudochattonella farcimen* (Dictyochophyceae). Protist *163*, 143–161.

Dorantes-Aranda, J.J., Nichols, P.D., Waite, T.D., Hallegraeff, G.M. (2013). Strain variability in fatty acid composition of *Chatonella marina* (Raphidophycea) and its relation to differing ichtyhotoxicity toward rainbow trout gill cells. J. Phycol. *49*, 427–438.

Eikrem, W., Edvardsen, B., Throndsen, J. (2009). Renaming *Verrucophora farcimen* Eikrem, Edvardsen et Throndsen. Phycol. Res. *57*, 170.

Edvardsen, B., Eikrem,W., Shalchian-Tabrizi, K., Riisberg, I., Johnsen, G., Naustvoll, L., Throndsen , J. (2007). *Verrucophora farcimen* gen. et. sp. nov. (Dictyochophyceae, Heterokonta) – A bloom-forming ichthytoxic flagellate from the Skagerrak, Norway. J. Phycol. *43*, 1054–1070.

Edvardsen, B., Eikrem, W., Throndsen, J., Sáezc , A.G., Probert, I., Medlin, L.K (2011) Ribosomal DNA phylogenies and a morphological revision provide the basis for a revised taxonomy of the Prymnesiales (Haptophyta). Europ. J. Phycol. *46(3)*, 202-228.

Food and Agriculture Organization of the United Nations. (2012). The state of world fisheries and aquaculture in 2012. (Rome, Italy: United Nations).

Ferreira, J.G., Andersen, J.H., Borja, A., Bricker, S.B., Camp, J., Cardoso da Silva, M., Garcés, E., Heiskanen, A.S.,Humborg, C.,Ignatiades, L., Lancelot, C., Menesguen, A., Tett, P., Hoepffnerm, N., Claussen, U. (2011). Overview of eutrophication indicators to assess environmental status within the European Marine Strategy Framework Directive. Estuar. Coast Shelf S. *93*, 117-131

Fryxell, G.A., Hasle, G.R. (2003). Taxonomy of harmful diatoms. In Manual on Harmful Marine Microalgae. Monographs on Oceanographic Methodology. Hallegraeff, G., Anderson, D.M, Cembella A.D., eds. (Paris, France: UNESCO Publishing), pp. 465-509.

Glibert, P., ed. Global ecology and oceanography of harmful algal blooms in eutrophic systems. (2006). (Paris, France and Baltimore,U.S.A.: Intergovernmental Oceanographic Commission-UNESCO and Scientific Committee on Oceanic Research).

Gentien, P., Reguera, B., Yamazaki, H., Fernand, L., Berdalet, E., Raine, R., eds. (2008). Global ecology and oceanography of harmful algal blooms: HABs in stratified systems. (Paris, France and Baltimore,U.S.A.: Intergovernmental Oceanographic Commission-UNESCO and Scientific Committee on Oceanic Research).

Glibert, P.M., Burkholder, J.M., Graneli, E., Anderson, D.M. (2008). Advances and insights in the complex relationships between eutrophication and HABs: Preface to the special issue. Harmful Algae *8*, 1-2.

Gómez, F. (2012). A checklist and classification of living dinoflagellates (Dinoflagellata, Alveolata). Cicimar Oceánides 27, 65-140.

Guiry, M.D., Guiry, G.M. (2015). AlgaeBase. World-wide electronic publication, National University of Ireland, Galway. http://www.algaebase.org; accessed at 14th November 2013.

Hallegraeff, G.M. (2003). Harmful algal blooms: a global overview. In Manual on Harmful Marine Microalgae. Monographs on Oceanographic Methodology. Hallegraeff, G., Anderson, D.M, Cembella A.D., eds. (Paris, France: UNESCO Publishing), pp.25-49.

Hallegraeff, G.M. (2010). Ocean climate change, phytoplankton community responses, and harmful algal blooms: a formidable predictive challenge. J. Phycol. *46*, 220-235.

Hallegraeff, G.M., Blackburn, S.I., Doblin, M.A., Bolch, C.J.S. (2012). Global toxicology, ecophysiology and population relationships of the chainforming PST dinoflagellate *Gymnodinium catenatum.* Harmful Algae *14*, 130-143.

Hallegraeff, G.M., Hara, T. (2003). Taxonomy of harmful marine raphidophytes. In Manual on Harmful Marine Microalgae. Monographs on Oceanographic Methodology. Hallegraeff, G., Anderson, D.M, Cembella A.D., eds. (Paris, France: UNESCO Publishing), pp. 511-522

Hays, G.C., Richardson, A.J., Robinson, C. (2005) Climate change and marine plankton. Trends Ecol. Evol. *20*, 337-344.

Hayward, P., Nelson-Smith, T., Shields, C. (1996). Seashore of Britain and Europe (London, U.K.: HarperCollins).

Hinder, S.L, Hays, G.C, Brooks, C.J, Davies, A.P., Edwards, M., Walne, A.W, Gravenor, M.B. 2011. Toxic marine microalgae and shellfish poisoning in the British isles: history, review of epidemiology, and future implications. Environmental Health-Glob. *10*, 54.

Heisler J., Glibert, P.M. Burkholder, J.M. Anderson, D.M. Cochlan, W. Dennison, W.C. Dortch, Q. Gobler, C.J. Heil, C.A. Humphries, E. Lewitus, A. Magnien, R. Marshall, H.G. Sellner, K. Stockwell, D.A. Stoecker, D.K. Suddleson, M. (2008). Eutrophication and harmful algal blooms: A scientific consensus, Harmful Algae *8*, 3-13.

Hess, P., Abadie, E., Hervé, F., Berteaux, T., Séchet, V., Aráoz, R., Molgó, J., Zakarian, A., Sibat, M., Rundberget, T., Miles, C.O., Amzil, Z. (2013). Pinnatoxin G is responsible for atypical toxicity in mussels (*Mytilus galloprovincialis*) and clams (*Venerupis decussata*) from Ingril, a French Mediterranean lagoon. Toxicon *75*, 16-26.

Hosoi-Tanabe, S., Honda, D., Fukaya, S., Otake, I., Inagaki, Y., Sako, Y. (2007). Proposal of *Pseudochattonella verruculosa* gen. nov., comb. nov. (Dictyochophyceae) for a formar raphidophycean alga *Chattonella verruculosa*, based on 18S rDNA phylogeny and ultrastructural characteristics. Phycol. Res. *55*, 185–192.

Imai, I., Yamaguchi, M., (2012). Life cycle, physiology, ecology and red tide occurrences of the fish-killing raphidophyte *Chattonella*. Harmful Algae *14*, 46–70.

James, K.J., Furey, A., Lehane, M., Ramstad, H., Aune, T., Hovgaard, P., Morris, S., Higman, W., Satake, M., Yasumuto, T. (2002). First evidence of an extensive northern European distribution of azaspiracid poisoning (AZP) toxins in shellfish. Toxicon *40*, 900-915.

James, K.J., Moroney, C., Roden, C., Satake, M., Yasumoto, T., Lehane, M., Furey, A. (2003). Ubiquitous 'benign' alga emerges as the cause of shellfish contamination responsible for the human toxic syndrom, azasoiracid poisoning. Toxicon *41*, 145-151.

James, K.J., Gillman, M., Amandi, M.F., López-Rivera, A., Puente, P.F., Lehane, M., Mitrovic, S., Furey, A. (2005). Amnesic shellfish poisoning in bivalve molluscs in Ireland. Toxicon *46*, 852-858.

Jauffrais, T., Marcaillou, C., Herrenknecht, C., Truquet, P., Séchet, V., Nicolau, E., Tillmann, U., Hess. P. (2012). Azaspiracid accumulation, detoxification and biotransformation in blue mussels (*Mytilus edulis*) experimentally fed *Azadinium spinosum*. Toxicon *60*, 582-595.

Kent, M.L., Whytel, J.N.C., LaTrace, C. (1995). Gill lesions and mortality in seawater pen-reared Atlantic salmon *Salmo salar* associated with dense bloom of *Skeletonema costatum* and *Thalassiosira* species. Dis. Aquat. Org. *22*, 77-81.

Krock, B., Tillmann, U., John, U., Cembella, A. (2009). Characterization of azaspiracids in plankton size-fractions and isolations of an azaspiracid-producing dinoflagellate from the North Sea. Harmful Algae *8*, 254-263.

Kudela, R.M., Gobler, C.J. (2012). Harmful dinoflagellate blooms caused by *Cochlodinium* sp.: Global expansion and ecological strategies facilitating bloom formation. Harmful Algae *14*, 71–86.

Kwok, C.T., van de Merwe, J.P, Chiu, J.M.Y, Wub, R.S.S. (2012). Antioxidant responses and lipid peroxidation in gills and hepatopancreas of the mussel *Pema viridis* upon exposure to the red-tide organism *Chattonella marina* and hydrogen peroxide. Harmful Algae *13*, 40–46.

Larsen, A. (1999). *Prymnesium parvum* and *P. patelliferum* (Haptophyta)- one species. Phycologia *38*, 541-543.

Lefebvre, K.A., Robertson, A. (2010). Domoic acid and human exposure risks: A review. Toxicon *56*, 218-230.

Lindholm, T., Öhman, P., Kurki-Helasmo, K., Koncaid, B., Meriluoto, J. (1999). Toxic algae and fish mortality in a brackish-water lake in Åland, SW Finland.Hydrobiologia *397*, 109-120.

Lømsland, E., Johnsen, T., Eikrem, W. (2010). Observations on *Chattonella globosa* in Norwegian coastal waters. Are *Chattonella globosa* and *Dictyocha fibula* one species? Paper presented at ICHA 14th Conference Proceedings. (Crete, Greece: International Society for the Study of Harmful Algae)

MacKenzie, L., Todd, K. (2002). *Alexandrium camurascutulum* sp. nov. (Dinophyceae): a new dinoflagellate species from New Zealand. Harmul Algae *1*, 295-300.

Magdalena, A.B., Lehane, M., Krys, S., Fernádez, M.L., Furey, A., James, K.J. (2003). The first identification of azaspiracids in shellfish from France and Spain. Toxicon *42*, 105-108.

Mann, D.G., Vanormelingen, P. (2013). An inordinate fondness? The number, distributions and origins of diatom species. J. Euk. Microbiol. *60*, 414-420.

McCarron, P., Rourke, W.A., Hardstaff, W., Pooley, B., Quilliam, M.A. (2012). Identification of pinnatoxins and discovery of their fatty acid ester metabolites in mussels (*Mytilus edulis*) from eastern Canada. J. Agric. Food Chem. *60*, 1437-46.

McNabb, P.S., McCoubrey, D.J., Rhodes, L., Smith, K., Selwood, A.I., van Ginkel, R., MacKenzie, A.L., Munday, R., Holland, P.T. (2012). New perspectives on biotoxin detection in Rangaunu Harbour, New Zealand arising from the discovery of pinnatoxins. Harmful Algae *13*, 34-39.

Michaloudi, E., Moustaka-Gouni, M., Gkelis, S., Pantelidakis, K. (2009). Plankton community structure during an ecosystem disruptive algal bloom of *Prymnesium parvum*. J. Plankt. Res. *31*, 301-309.

Moestrup, Ø. (1994). Economic aspects: 'blooms', nuisance, species and toxins. In The Haptophyte Algae, Systematics Association, Special Volume 51, Green, J.C., Leadbeater, B.S.C., eds. (Oxford , U.K.: Clarendon Press), pp. 265-284.

Moestrup, Ø., Akselman, R., Cronberg, G., Elbraechter, M., Fraga, S., Halim, Y., Hansen, G. Hoppenrath, M., Larsen, J., Lundholm, N., Nguyen, L.N., Zingone, A., eds. (2013) IOC-UNESCO Taxonomic Reference List of Harmful Microalgae. Available online at http://www.marinespecies.org/HAB. Accessed at 14th November 2013.

Moestrup, Ø., Thomsen, H.A. (2003) Taxonomy of toxic haptophytes (prymnesiophytes). In Manual on Harmful Marine Microalgae. Monographs on Oceanographic Methodology. Hallegraeff, G., Anderson, D.M, Cembella A.D., eds. (Paris, France: UNESCO Publishing), pp. 433-464.

Montresor, M., John, U., Beran, A., Medlin, L.K. (2004). *Alexandrium tamtum* sp. nov. (Dinophyceae): A new nontoxic species in the genus *Alexandrium*. J. Phycol. *40*, 398-411.

Munday, R., Selwood, A.I., Rhodes, L. (2012). Acute toxicity of pinnatoxins E, F and G to mice. Toxicon *60*, 9950999

Naustvoll, L.J. (2006). NOBANIS – Invasive alien species fact sheet - *Chattonella* aff. *verruculosa*. From Online Database of the North European and Baltic Network on Invasive Alien Species. (North European and Baltic Network on Invasive Alien Species, NOBANIS). Date of access 30/10/2013.

O'Neil, J.M., Davis, T.W., Burford, M.A., Gobler, C.J. (2012). The rise of harmful cyanobacteria blooms: The potential roles of eutrophication and climate change. Harmful Algae *14*, 313–334

Negri, A.P., Bunter, O., Jones, B., Llewellyn, L. (2004). Effects of the bloom-forming alga *Trichodesmium erythraeum* on the pearl oyster *Pinctada maxima*. Aquaculture *232*, 91-102.

Nézan, E., Chomérat, N. (2011). *Vulcanodinium rugosum* gen. et sp. nov. (Dinophyceae), un nouveau dinoflagellé marin de la côte méditerranéenne française. Cryptogamie Algol. *32*, 3-18.

Paredes, I., Rietjens, I.M.C.M, Vietes, J.M., Cabado, A.G. (2011). Update of risk assessments of main marine biotoxins in the European Union. Toxicon *58*, 336-354.

Parsons, M., L., Aligizaki, K., Bottein, M.Y.D., Fraga, S., Morton, S.L., Penna, A., Rhodes, L. (2012). *Ganbierdiscus* and *Ostreopsis*: Reassessment of the state of knowledge of their taxonomy, geography, ecophysiology, and toxicology. Harmful Algae *14*, 107-129.

Paz, B., Daranas, A.H., Norte, M., Riobó, P., Franco, J.M., Fernandes, J.J. (2008). a Group of Marine Polyether Toxins: an Overview. Mar. Drugs *6*, 73-102.

Pettersson, L.H., Pozdnyakov, D. (2013). Monitoring of Harmful Algal Blooms. (Berlin, Germany: Springer-Verlag).

Pezzolesi, L., Cucchiari, E., Guerrini, F., Pasteris, A., Galletti, P., Tagliavini, E., Totti, C., Pistocchi, R. (2010). Toxicity evaluation of *Fibrocapsa japonica* from the Northern Adriatic Sea through a chemical and toxicological approach. Harmful Algae *9*, 504–514.

Pitcher, G.C. Figueiras, F.G. Hickey, B.M. Moita. M.T. (2010). The physical oceanography of upwelling systems and the development of harmful algal blooms. Prog. Oceanogr. *85*, 5–32.

Pitcher, G., Moita, T., Trainer, V., Kudel, R., Figueiras, P. Probyn, T. eds.(2005). Global ecology and oceanography of harmful algal blooms: HABS in upwelling systems. (Paris, France and Baltimore,U.S.A.: Intergovernmental Oceanographic Commission-UNESCO and Scientific Committee on Oceanic Research).

Place, A.R., Bowers, H.A., Bachvaroff, T.R., Adolf, J.E., Deeds, J.R., Sheng, J. (2012). *Karlodinium veneficum* - The little dinoflagellate with a big bite. Harmful Algae *14*, 179-195.

Raine, R., McDermott, G., Silke, J., Lyons, K., Nolan, G., Cusack, C. (2010). A simple short range model for the prediction of harmful algal events in the bays of southwestern Ireland. J. Mar. Syst. *83*, 150-157.

Raine,R. (2013). Review of the biophysical interactions relevant to the promotion of HABs in stratified systems: The case study of Ireland. Deep-Sea Res. Pt II *101*, 21-31.

Reguera, B., Velo-Suárez, L., Raine, R., Park, M.G. (2012). Harmful *Dinophysis* species: A review. Harmful Algae *14*, 87-106.

Rensel, J.E. (1993). Severe blood hypoxia of Atlantic Salmon (*Salho salar*) exposed to the marine diatom *Chaetoceros concavicornis*. In Toxic Phytoplankton Blooms in the Sea, Smayda, T.S., Shimuizu, Y., eds. (New York, U.S.A.: Elsevier), pp. 623-625.

Riisberg, I., Edvardsen, B. (2008). Genetic variation in bloom-forming ichthyotoxic *Pseudochattonella* species (Dictyochophyceae, Heterokonta) using nuclear, mitochondrial and plastid DNA sequence data. Europ. J Phycol. *43(4),* 413-422.

Rijssel, M.v., Alderkamp,A.C., Njestgaard, J.C., Saxhin, A.F., Verity, P.G. (2007). Heamolytic activity of live *Phaeocystis pouchetii* during mesocosm blooms. Biogeochemistry *83*, 189-200.

Rines, J.E.B., McFarland, M.N., Donaghay, P.L., Sullivan, J.M. (2010). Thin layers and species-specific characterization of the phytoplankton community in Monterey Bay, California, U.S.A. Cont. Shelf Res *30*, 66-80.

Rhodes, L., Smith, K., Selwood, A.,McNabb, P., van Ginkel, R., Holland, P., Munday, R. (2010). Production of pinnatoxins by a peridinoid dinoflagellate isolated from Northland, New Zealand. Harmful Algae *9*, 384–389

Rundberget, T., Aasen, J.A.B., Selwood, A.L., Miles, C.O. (2011). Pinnatoxins and spirolides in Norwegian blue mussels and seawater. Toxicon *58*, 700–711

Simó, R. (2001). Production of atmospheric sulfur by oceanic plankton: biogeochemical, ecological and evolutionary links. Trends Ecol. Evol. *16(6)*, 287-294.

Sipiä, V., Kankaanpää, H., Peltonen, H., Vinnic, M., Meriluoto, J. (2007). Transfer of nodularin to three-spined stickleback (*Gasterosteus aculeatus* L.), herring (*Clupea harengus* L.), and salmon (*Salmo salar* L.) in the northern Baltic Sea. Ecotox. Environ. Safety *66*, 421–425.

Skjelbred, B., Horsberg, T.E., Tollefsen, K.E., Andersen, T., Edvardsen, B. (2011). Toxicity of the ichthyotoxic marine flagellate *Pseudochattonella* (Dictyochophyceae, Heterokonta) assessed by six bioassays. Harmful Algae *10*, 144–154.

Smayda, T.J. (2006). Harmful algal bloom communities in Scottish coastal waters: Relationship to fish farming and regional comparisons – a review. (Scotland, United Kingdom: Scottish Executive Environment Group).

Smith, K.F., Rhodes, L.L., Suda, S., Selwood, A.I. (2011). A dinoflagellate producer of pinnatoxin G, isolated from sub-tropical Japanese waters. Harmful Algae *10*, 702–705

Speare, D.J., Brackett, J., Fergusen, H.M. (1989). Sequential pahtology of the gills of Coho salmon with a combined diatom and microsporidian gill infection. Can. Vet. J. *30*, 571-575.

Suikkanen, S., Kremp, A., Hautala, H., Krock, B. (2013). Paralytic shellfish toxins or spirolides? The role of environmental and genetic factors in toxin production of the *Alexandrium ostenfeldii* complex. Harmful Algae *26*, 52-59.

Sullivan, J. M., McManus, M.A., Cheriton, O.M., Benoit-Bird, K.J., Goodman, L., Wang, Z., Ryan, J.P., Stacey, M., Holliday, D.V., Greenlaw, C.F., Moline, M.A., McFarland, M. (2010). Layered organization in the coastal ocean: an introduction to thin layers and the loco project. Cont. Shelf Res. *30*, 1-6.

Tangen, K. (1999). Skadelig plankton i fiskeoppdrett. In Fiskehelse og fiskesykdommer, Poppe, T., ed., (Oslo, Norway: Universitetsforlaget), pp.252-265. (In Norwegian)

Taylor, F.J.R., Fukuyo, Y., Larsen, J., Hallegraeff, G.M. (2003). Taxonomy of harmful dinoflagellates. In Manual on Harmful Marine Microalgae. Monographs on Oceanographic Methodology. Hallegraeff, G., Anderson, D.M, Cembella A.D., eds. (Paris , France: UNESCO Publishing), pp. 389-432.

Tillmann, U., Elbächter, M., Krock, B., John, U., Cembella, A. (2009). *Azadinium spinosum* gen. et sp. nov. (Dinophyceae) identified as a primary producer of azaspiracid toxins. Eur. J. Phycol. *44(1)*, 63-79.

Torgersen, T., Bremnes, N.B., Rundberget, T., Aune, T. (2008). Structural confirmation and occurrence of azaspiracids in Scandinavian brown crabs (*Cancer pagurus*). Toxicon *51*, 93-101.

Touzet, N., Franco, J.M., Raine, R. (2008). Morphogenetic diversity and biotoxin composition of *Alexandrium* (Dinophyceae) in Irish coastal waters. Harmful Algae *7*, 782-797

Trainer, V.L., Pitcher G.C., Reguera B., Smayda T.J. (2010). The distribution and impacts of harmful algal bloom species in eastern boundary upwelling systems. Progr. Oceanogr. *85*, 33-52.

Treasurer, J.W., Hannah, F., Cox, D. (2003). Impact of a phytoplankton bloom on mortalities and feeding response of farmed Atlantic salmon, *Salmo salar*, in west Scotland. Aquaculture *218*, 103- 113.

Uemura, D., Chou, T., Haino, T., Nagatsu, A., Fukuzawa,S., Zheng, S., Chen, H. (1995). Pinnatoxin A: A toxic amphoteric macrocycle from the Okinawan bivalve *Pinna muricata*. J. Am. Chem. Soc. *117*, 1155-1156.

Usup, G., Ahmad, A., Matsuoka, K., Lim, P.T., Leaw, C.P. (2012). Biology, ecology and bloom dynamics of the toxic marine dinoflagellate *Pyrodinim bahamense*. Harmful Algae *14*, 301-312.

Vasas, G., M-Hamvas, M., Borics, G., Gonda, S., Máthé, C, Jámbrik, K., Nagy, Z.L. (2012). Occurrence of toxic *Prymnesium parvum* blooms with high protease activity is related to fish moratlity in Hungarian ponds. Harmful Algae *17*, 102-110.

Worth, G.K., MacLean, J.L., Price, M.J. (1975). Paralytic shellfish poisoning in Papau New Guinea, 1972. Pacific Science *29(1)*, 1-5.

Yang, C.Z., Albright, L.J. (1994). The harmful phytoplankter *Chaetoceros concavicornis* causes high mortalities and leukopenia in Chinook salmon (*Oncorhynchus tshawytscha*) and Coho salmon (*Oncorhynchus kisutch*). Can. J. Fish. Aquat. Sci. *51(11)*, 2493-2500.

Yasumoto, T. (1998). Fish poisoning due to toxins of microalgal origins in the Pacific. Toxicon *36(11)*, 1515-1518.

Zingone, A., Siano, R., D'Alelio D.,Sarno, D. (2006). Potentially toxic and harmful microalgae from coastal waters of the Campania region (Tyrrhenian Sea, Mediterranean Sea). Harmful Algae *5*, 321–337.

Other Publications of Interest

Flores, E. and Herrero, A. (2014). The Cell Biology of Cyanobacteria. Caister Academic Press, U.K. ISBN: 978-1-908230-38-6

Herold, K.E. and Rasooly, A. (2009). Lab-on-a-Chip Technology: Fabrication and Microfluidics. Caister Academic Press, U.K. ISBN: 978-1-904455-46-2

Herold, K.E. and Rasooly, A. (2009) Lab-on-a-Chip Technology: Biomolecular Separation and Analysis. Caister Academic Press, U.K. ISBN: 978-1-904455-47-9

Wilkinson, M.G. (2015). Flow Cytometry in Microbiology: Technology and Applications. Caister Academic Press, U.K. ISBN: 978-1-910190-11-1

Xuefeng, L. (2014). Biofuels: From Microbes to Molecules. Caister Academic Press, U.K. ISBN: 978-1-908230-45-4

5

Emiliania huxleyi in the Genomic Era

Bethan M. Jones[1]*, Mónica Rouco[2], M. Débora Iglesias-Rodriguez[3] and Kimberly H. Halsey[4]

[1]Department of Botany and Plant Pathology, Cordley Hall 2082, Oregon State University, Corvallis, OR 97331-2902, U.S.A. bethan.jones@science.oregonstate.edu
[2]Lamont-Doherty Earth Observatory, Columbia University, NY, U.S.A. molina@ldeo.columbia.edu
[3]Department of Ecology, Evolution and Marine Biology, University of California Santa Barbara, CA, U.S.A. iglesias@lifesci.ucsb.edu
[4]Department of Microbiology, Oregon State University, Corvallis, OR, U.S.A. halseyk@science.oregonstate.edu
*Author for correspondence.

Contents

Abstract

Emiliania huxleyi is a globally abundant coccolithophore species that plays a significant biogeochemical role in modern oceans. The species has become an important model organism, with extensive research conducted on its responses to carbon chemistry, light regimes and nutrient dynamics. In the years since the seminal review on *E. huxleyi* by Paasche (2002), the field of microbial oceanography has been revolutionized by the increasing application of genomic and post-genomic technology. Paasche intended his review as a starting point for future molecular studies on coccolithophores, a field that in 2002 was described as "barely in its infancy". This chapter revisits Paasche (2002), using the review as a framework to describe how our knowledge of *E. huxleyi* has progressed since these tools have become more commonly applied. The same key areas that Paasche addressed are focused upon, namely how the morphology, diversity, life cycle and physiology of this species affects calcification. This is used to provide avenues for future research and suggest directions to help answer many of the questions still posed regarding the biology of this globally important yet seemingly enigmatic organism.

Introduction

Coccolithophores are single-celled, eukaryotic marine phytoplankton distinguished by their ability to produce calcite plates (coccoliths) that cover the exterior of the cell. In the modern ocean, coccolithophores are one of the main sources of calcite ($CaCO_3$) to deep-sea sediments (Milliman 1993; Westbroek et al., 1993) and may fix up to 20% of carbon in some systems (Poulton et al., 2007). *Emiliania huxleyi* (Lohmann) Hay et Mohler, a species that first emerged between 265,000 to 290,000 years ago (Thierstein et al., 1977; Raffi et al., 2006), is the most abundant coccolithophorid in modern oceans and often dominates counts from oceanic sediment traps (Beaufort et al., 2007). *E. huxleyi* has a wide distribution, forming between 30-70% of mid-latitude and 10-20% of low-latitude coccolithophore floras (Westbroek et al., 1989). The species also forms extensive subpolar blooms (Brown and Yoder, 1994; Iglesias-Rodriguez et al., 2002). It is this ecological and biogeochemical importance, as well as the ease with which *E. huxleyi* can be maintained in culture, that have made the species a suitable model organism for studies on coccolithophore biology (Westbroek et al., 1993). Consequently, *E. huxleyi* ecophysiology has been studied intensively and was the focus of an extensive review on the species by Paasche in 2002. Since then, a wealth of knowledge has emerged warranting a re-evaluation of the biology and physiology of this important organism. With this in mind, it is intended that this review is a companion text to Paasche (2002). However, befitting a review on such a biogeochemically important species, not all aspects can be covered in a book chapter. Similar to Paasche, fields such as *E. huxleyi*-virus interactions, alkenones and DMS production are not discussed here.

Genetic and morphological diversity

Evidence was already beginning to come to light that *E. huxleyi* was a morphologically and genetically diverse species in 2002, when Paasche compiled his review. This issue of variability likely affects most factors regarding the responses of this species to differing environmental dynamics and extensive efforts have been directed to understanding intraspecific diversity in recent years. A plethora of genomic studies and the recent publication of the *E. huxleyi* genome (Read et al., 2013) have helped us appreciate that this species may be an extremely complex driver of oceanic biogeochemistry.

Morphotype diversity

At present, seven different morphotypes of *E. huxleyi* are typically recognized; A, B, B/C, C, O, R and corona (Figure 5.1), based on immunological, biochemical, geographic, physiological, genetic and morphometric differences (e.g. van Bleijswijk et al., 1991; Young and Westbroek, 1991; Young et al., 2003; Hagino et al., 2005; Hagino et al., 2011). Some authors have raised these morphotypes to distinct varieties (var. *huxleyi* [type A], var. *pujosae* [type B], var. *kleijniae* [type C], Medlin et al., 1996; var. *aurorae* [type B/C], Cook et al., 2011). Type D is an additional morphotype occasionally mentioned in the literature (Findlay and Girardeau et al., 2000; Mohan et al., 2008) but it is likely that this exists as an artifact caused by *in situ* seawater chemistry, collection techniques, microscopy or fixation.

Recent evidence supports splitting morphotypes of *E. huxleyi* into two groups: A (composed of types A, R and corona) and B (composed of types B, B/C, C and O) based on shape and degree of calcification (Young, pers. comm.). Group B cells are all lightly calcifying and include the recently defined type O that possesses coccoliths similar to

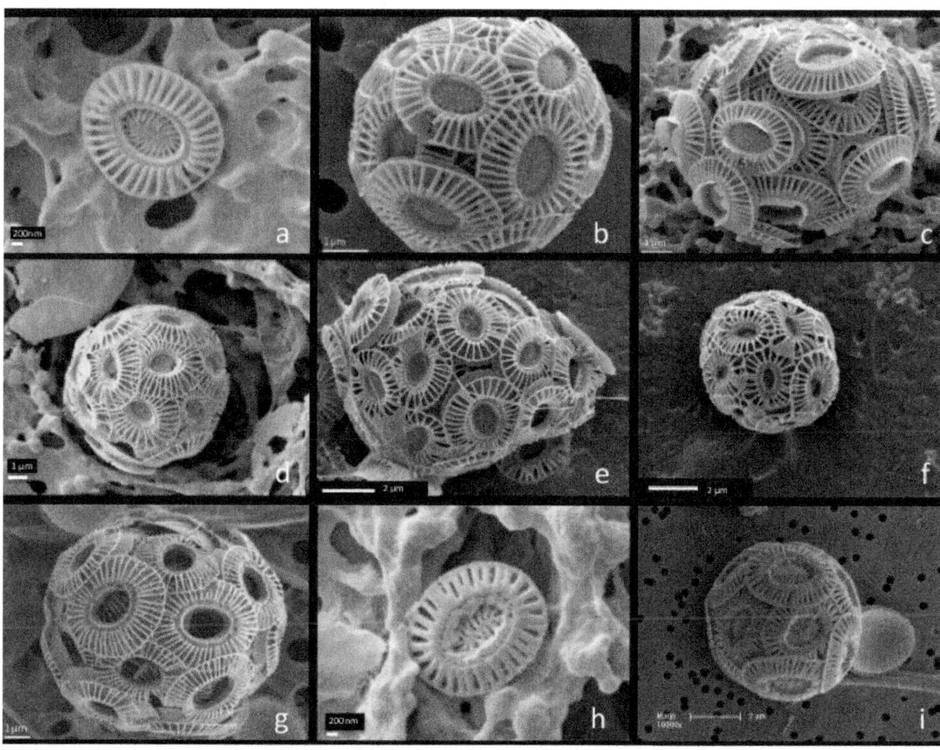

Figure 5.1. SEM images showing different morphotypes of *Emiliania huxleyi*, plate a) morphotype A coccolith (isolated from waters around Svalbard in 2010), b) morphotype A coccosphere (unknown origin), c) *E. huxleyi* var. corona (Canary Islands, Spain 1992), d) morphotype B coccosphere (Svalbard, 2010), e) type B/C coccosphere (Patagonian Shelf, 2008), f) type C coccosphere (Patagonian Shelf, 2008), g) morphotype O coccosphere (unknown origin), h) morphotype R coccolith (Svalbard, 2010), i) morphotype R coccosphere (strain NZEH, isolated from Bounty Bay, New Zealand, 1992). Image credits: b), c) and g) Jeremy Young; e) and f) Alex Poulton; remainder Bethan Jones, unpublished.

type B but with an open central area (Hagino et al., 2011). Type O is typically easier to classify in culture but in the field it is more difficult to discern from other morphotypes since it often co-occurs with type B, whilst types B/C and C may also have open central areas. This means some early field distribution reports may require reappraisal. There is some variation in calcification within group A, which includes R, a distinct morphotype that resembles cells of the genus *Reticulofenestra* (Young et al., 2003) and corona, which is similar to A but with a protruding surface. Morphotype A has also been observed as "overcalcified" in some waters, bearing coccoliths with a closed central area (Smith et al., 2012).

It is becoming clear that different morphotypes of *E. huxleyi* may have differential distribution patterns. B appears to be a Northern hemisphere type that is particularly common in the North Sea and around the British Isles (van Bleijswijk et al., 1991) but it is not known if this morphotype is more geographically extensive (Hagino et al., 2011). Type A is known to have a global distribution, appearing in samples from both hemispheres (e.g. Cubillos et al., 2007; Beaufort et al., 2011; Poulton et al., 2011; Smith et al., 2011). Meanwhile, B/C (*sensu* Young et al., 2003) has so far only observed in sub-polar (Hagino et al., 2005) and Southern hemisphere waters (Finlay and Girardeau, 2000; Mohan et al., 2008; Cubillos et al., 2007; Henderiks et al., 2012; Hinz et al., 2012) with an apparent preference for nutrient rich but colder (<10°C) waters (Poulton et al., 2011). Moreover, Southern Ocean B/C strains are genetically distinct from Southern Ocean type A isolates and possess a different pigment profile (Cook et al., 2011), suggesting differences in morphotype ecophysiology. Distinct hydrographic preferences (Beaufort et al., 2011), additional genetic differences between type A and B (Schroeder et al., 2005), and the presence of multiple unique genotypes of type A (Schroeder et al., 2005), B/C and R (Cook et al., 2011) indicate that *E. huxleyi* is a pseudo-cryptic species complex harboring extensive genetic variability.

Genetic diversity

Since the 2002 review, the application of genomic tools and genome sequencing has become seemingly commonplace. High levels of intraspecific diversity originally proposed by Brand (1982), and suggested by RAPD fingerprinting (Medlin et al., 1996), have now been confirmed using a variety of methods. For example, Iglesias-Rodriguez and co-authors (2002) conducted AFLP on nine strains to suggest that there are distinct Northern and Southern hemisphere populations. Hagino et al. (2011) investigated the genetic variability of 39 strains, isolated from a diversity of ocean basins, using mitochondrial *cox1b* and *atp4* genes to find two clades differentiated by water temperature and phosphate concentration. Denaturing gradient gel electrophoresis using the calcium binding *GPA* genes (see section "Genomics of calcification") suggested that *E. huxleyi* blooms in the North Sea were dynamic and diverse assemblages (Martinez et al., 2012). Meanwhile, microsatellite analysis on 85 global isolates (Iglesias-Rodriguez et al., 2006) and 273 Southern hemisphere strains (Cook et al., 2013) indicated that *E. huxleyi* was a species complex composed of numerous genetically diverse phenotypes selected by environmental pressures.

The recent publication of the *E. huxleyi* reference genome of strain CCMP1516 accompanied with the sequences of 13 additional strains confirmed the presence of vast levels of diversity at the molecular level. This suggested that the species actually exists as a pan-genomic complex, with some strains only exhibiting 54-77% similarity to the reference genome (Read et al., 2013). Moreover, the presence and absence of genes between strains was exhibited, in addition to variable copy numbers of genes involved in nutrient acquisition and metabolism. This lead Read and co-authors to make the same conclusion as Paasche; namely that a single strain will never be a perfect representative for all strains. More so than ever, *E. huxleyi* should be considered a plastic, rapidly adapting and resilient entity. Indeed, Kegel and co-authors (2013) recently discovered that gene content sometimes varies by up to 30% in different strains of *E. huxleyi*, leading the authors to conclude that the species is undergoing a rapid evolutionary radiation and should be described as a species-morphotype complex. It is likely that this strain variability underlies many aspects of the various reported responses of coccolithophores to ocean acidification (Langer et al., 2009; Ridgwell et al, 2009) and other environmental variables (see section "Ecophysiology and responses to environmental change"). Diversity observed in other coccolithophore species (Geisen et al., 2004; Quinn et al., 2004) suggests that this may be a common characteristic for marine eukaryotic phytoplankton and is a feature that has been frequently recognized in bacteria (Tettelin et al., 2008).

Ultimately, it remains to be seen whether variations in morphotype and the underlying genetic diversity of *E. huxleyi* have biogeochemical ramifications. Although there is considerable overlap in the size of coccoliths produced between morphotypes (Poulton et al., 2011), evidence is beginning to come to light that different morphotypes can have distinct impacts on regional calcification rates. For example, blooms of B/C morphotype cells have significantly lower $CaCO_3$ production rates compared to blooms composed of other morphotypes (Poulton et al., 2013), theoretically affecting surface-deep sea CO_2 and $CaCO_3$ fluxes. Additionally, the levels of genetic diversity exhibited by the *E. huxleyi* species group suggest that it is an exceptionally complex driver of elemental cycling. For example, genes coding for proteins involved in the uptake or assimilation of nutrients, such as phosphate transporters, alkaline phosphatase and nitrate reductase, amongst others, did not show a uniform distribution between strains (Read et al., 2013). The functional implications of this substantial diversity have also been observed in a recent enzymatic activity study (Reid et al., 2011). Therefore, the questions now needing to be taken into consideration are: 1) what is the underlying cause of this level of variability, 2) can these functional patterns be linked to niche partitioning within the species group, and 3) what are the biogeochemical implications of this intra-specific diversity?

Life cycle considerations

Although Paasche (2002) already reviewed the existence of different types of *E. huxleyi* cells associated with life cycle stages and the potential biogeochemical differences between them, life cycles have been a factor often neglected in both field and laboratory studies. However, life cycles in marine phytoplankton are beginning to be reconsidered (see von Dassow and Montresor, 2011). Three cell types are still recognized for *E. huxleyi*; C (coccolith-forming), N (naked) and S (scaly) but to our knowledge, there have been no further mention of amoeboid cells in the literature since the review of Paasche. As before, it is generally accepted that C cells are diploid adult stages, N cells are mutant diploid non-calcifiers and S cells are haploid, non-calcifying, motile gametes. It is now generally accepted that coccolithophores alternate asexuality by binary fission with sexual reproduction by syngamy (Klaveness, 1972; Green et al., 1996), although the latter process has not yet been observed. Anecdotally, Houdan et al. (2004) proposed that meiosis might relate to circadian rhythms or seasonality, whilst Frada et al. (2008) suggested that reactive oxygen species were responsible, however factors influencing *E. huxleyi* phase shifts remain unknown. Microsatellite analyses suggested that the species may regularly undergo sexual reproduction (Iglesias-Rodriguez et al., 2006; Cook et al., 2013). However, each stage is able to reproduce vegetatively in an indefinite manner, as confirmed by culture studies (Billard and Inouye, 2004), and haploid cells may persist at low (<0.1 cells mL^{-1}) levels throughout the year (Holligan et al., 1993). It is worth noting that the study by Holligan and co-authors (1993) is a rarity with respect to reporting haploid concentrations in the field.

So far, the distribution of S (haploid) cells in the environment is unknown, since they are difficult to differentiate from small flagellates and are not well preserved using classical techniques (von Dassow and Montresor, 2011) and most field studies focus on the diploid "C" stage cells. Following early attempts by Campbell et al. (1994) to detect haploid cells in the field using immunofluorescence, Frada et al. (2006) developed a method, combining polarized optical microscopy with fluorescent *in situ* hybridization, to enable the detection of both haploid and diploid coccolithophore cells. This was used to confirm that haploid cells of *E. huxleyi* were present within Norwegian fjords (Frada et al., 2012) but little else is known about the distribution of these cells in the environment.

It is likely that haploid and diploid cells of *E. huxleyi* represent phases that are evolutionary shaped to occupy specific niches in different environments. Unlike calcified diploid cells, uncalcified haploid cells are motile, have a stronger defense response to grazers (Kolb and Strom, 2013), are resistant to viral attack (Frada et al., 2008), and may also be more sensitive to light (Houdan et al., 2005) and high CO_2 (Fiorini et al., 2011). At the molecular level, succinct differences have also been observed between haploid and diploid cells of *E. huxleyi*. For example haploid cells exhibited higher transcriptional levels of genes associated with nitrogen and carbon metabolism, as well as vitamin synthesis when compared to their diploid counterparts (Rokitta et al., 2011). Expressed sequence tag analysis also indicated dramatic differentiation (only ~50% of transcript clusters shared) between the two cell phases, with haploid cells producing more transcripts associated with light sensing and signal transduction (von Dassow et al., 2009). Despite differentiation at the transcriptional level, differences in the activity of a series of enzymes (e.g. acid/alkali phosphodiesterase, acid/alkali phosphomonoesterase, EC1.1.1-type dehydrogenase, EC1.3.1-type dehydrogenase and carboxylesterase) were not observed between haploid and diploid phases of the same *E. huxleyi* strain (Reid et al., 2011). Expanded investigations are therefore required in order to investigate if transcriptional and functional patterns are uncoupled.

Recent evidence suggests that ploidy level may affect biogeochemistry. For example, haploid cells produce fewer transparent exopolymer particles (TEP) than diploid cells (Pedrotti et al., 2012; Van Oostende et al., 2013), but exude a higher concentration of dissolved carbohydrates (Van Oostende et al., 2013). This may relate to calcification status, since coccoliths have an organic coating that may be transformed to form TEP

(Godoi et al., 2009). A lower concentration of TEP in the haploid stage could alter export efficiency by preventing dissolved polysaccharides being coagulated and exported to depth (see Engel et al., 2004). While it is clear that cell stage could therefore have substantial biogeochemical implications, this requires further investigation. To better understand ocean carbon cycling, we must now turn out attention to consider when, where and why different cell phases arise.

Although mechanisms for phase shifts in *E. huxleyi* remain unknown, models have suggested that haploid-diploid lifecycles persist when niche differentiation and haploid resistance to competition is sufficient to compensate for the benefits of being diploid (namely higher survival and reproduction; Hughs and Otto, 1999). This suggests that *E. huxleyi* populations exist through an evolutionary stable strategy (*sensu* Maynard Smith and Price, 1973) and significant differences between haploid and diploid phases enables their persistence. Accordingly, diploid cells may be able to occupy numerous niches, since they have a richer transcriptome (von Dassow et al., 2009) but they lack additional benefits of motility, such as locomotion and ability to generate thin nutrient filaments *via* turbulence (Taylor and Stocker, 2012). In addition to carbon and nitrogen metabolic requirements that are putatively different (Rokitta et al., 2011), mobility in the haploid phase may enable these cells to exist in environments that are not as advantageous to their diploid counterparts. Additionally, the smaller size of haploid cells results in a greater surface area to volume ratio, enabling a greater uptake of nutrients and an enhanced competitive ability under low nutrient regimes. This leads to a possibility that they are able to exist in more oligotrophic regions than diploid cells (Qiu et al., 2012). The strategy of viral resistance in the haploid phase is also beneficial; since deleterious alleles are not masked, they can be purged from the population and successful new variants quickly fixed (Otto and Marks, 1996). This would enable haploid populations to adapt more rapidly than those that are diploid (Zeyl et al., 2003). However, it may be that in other environments, the benefits of the haploid phase are outweighed by the benefits of recombination and meiosis in the diploid cells. We suggest that the haploid stage of *E. huxleyi* may constitute a relatively persistent life stage and efforts need to become more focused on accounting for this in future research.

Coccoliths and calcification

Turning our attention back to the more frequently studied diploid (C) cells of *E. huxleyi*, a large section of the remainder of this chapter relates to how calcification is affected by environmental parameters. Since Paasche (2002), a wealth of genomic studies have attempted to understand the molecular mechanisms underlying calcification. Additionally, attempts have continued to address whether calcification acts as a carbon-concentrating mechanism for photosynthesis. To introduce or perhaps refresh the reader, we will now briefly re-review cytology as a pre-requisite to addressing the chemistry and genomics associated with calcification.

Cytology

The cellular mechanisms controlling the complex morphology of coccoliths and the variation in rates of production of $CaCO_3$ were a focus of attention for biologists and ecophysiologists over the last century, following the discovery of coccoliths by C.G. Ehrenberg (1795–1875) and their identification by T. Huxley and C. Emiliani. The cellular components involved in coccolithophore calcification have been investigated mainly in *E. huxleyi*, as well as in other species such as *Coccolithus pelagicus* and *Pleurocrysis carterae* (Manton and Leedale, 1969; van der Wal et al., 1987; Marsh, 1994; Taylor et al., 2007). Briefly, coccolithogenesis begins with the precipitation of $CaCO_3$ intracellularly, within the coccolith vesicle, a specialized compartment that seems to arise from merged membranes of Golgi origin in association with the nucleus (van der Wal et al., 1983). In some species (*E. huxleyi* and *Gephyrocapsa* sp.), a reticular body has been described associated with the distal side of the coccolith vesicle. These intracellular structures are also present in non-calcifying coccolithophore cells, apparently in a deformed state and less closely associated with the nucleus (Klaveness and Paasche, 1971; van der Wal et al., 1983). The synthesis of coccoliths involves the formation of an organic scale or baseplate followed by nucleation of $CaCO_3$ crystals (Westbroek et al., 1989; Marsh, 1999; Young et al., 1999). The construction and mechanisms controlling $CaCO_3$ crystal formation appear to be highly dependent on calcium binding proteins (Corstjens et al., 1998; see section "Genomics of calcification") and polysaccharides, which have been identified as tightly coupled with coccoliths (van der Wal et al., 1983; Marsh, 1994; Henriksen et al., 2004). Direct observations under the microscope have revealed that the formation of a coccolith is rapid, between one (Paasche, 1962) and 3.5 hours (Taylor et al., 2007) and its secretion involves considerable contractile activity to eject and position the coccolith on the cell surface (Taylor et al., 2007).

Inorganic carbon and calcium kinetics

Calcification is the biogenic precipitation of $CaCO_3$ from Ca^{2+} and CO_3^{2-} precursors. Given that *E. huxleyi* precipitates $CaCO_3$ intracellularly, cells require the uptake and delivery of Ca^{2+} and inorganic carbon to the site of $CaCO_3$ nucleation. Cells of *E. huxleyi* appear to import HCO_3^- into the cytosol and it is thought that HCO_3^- ions are also imported into the coccolith vesicle, possibly by a Cl^-/HCO_3^- antiport mechanism (Holtz et al., 2013; see section "Genomics of calcification" for a more thorough discussion of the potential transporters involved in calcification). Inside the coccolith vesicle and calcification fluid, the conversion of HCO_3^- into CO_3^{2-} is required in order to maintain a supply of inorganic carbon for $CaCO_3$ precipitation. In this process, H^+ are generated and these must be extruded in order to maintain elevated pH in the calcification fluid. A model proposing the activity of carbonic anhydrase (CA, an enzyme catalyzing the interconversion of HCO_3^-, CO_2 and H_2O) inside the coccolithophore vesicle, a Ca^{2+}/H^+ exchanger, and a supply of dissolved inorganic carbon (DIC) exclusively *via* CO_2 diffusion has been proposed as the most likely mechanism for calcification because CA and Ca^{2+}/H^+ exchangers have already been discovered in *E. huxleyi* (Holtz et al., 2013). Furthermore, this method is likely to be energetically advantageous given that the pH of the cytosol is significantly lower than that of the calcification fluid in *E. huxleyi* (Dixon et al., 1989; Nimer et al., 1994).

Many studies have focused on the mechanisms controlling the acquisition of substrates required for coccolith formation although perhaps fewer exist about the mineralogical characteristics of coccoliths. Mineralogical composition is important when trying to predict the thermodynamic stability and preservation of $CaCO_3$ based on not only its physiological controls, but also the chemical processes affecting dissolution. Indeed, it has been assumed that coccolithophores produce $CaCO_3$ only in the form of calcite;

however, evidence suggests that coccoliths can also be made of aragonite or vaterite (Wilbur and Watabe, 1963), two polymorphs of $CaCO_3$ with substantially higher susceptibilities to dissolution (Morse et al., 2006). This work has been disputed by Young et al. (1991), but the two studies used different strains suggesting that different subspecies/strains may display different mineralogies under varying environmental conditions.

Carbon sources in photosynthesis and biogeochemical implications

Neither photosynthesis nor calcification in *E. huxleyi* appears to be saturated at ambient concentrations of DIC in seawater (Herfort et al., 2002). The form of DIC used by ribulose-1,5-bisphosphate carboxylase/oxygenase (RuBisCO) during photosynthesis is CO_2 (Cooper et al., 1969). However, the concentration of CO_2 in seawater is only ~1% of the total DIC compared to bicarbonate, with makes up ~90% of DIC. To overcome this limitation, most photosynthetic organisms have developed a carbon concentrating mechanism (CCM), by which either CO_2 or HCO_3^- is transported actively across membranes. However, coccolithophores are considered to have generally inefficient CCMs (Badger et al., 1998; Reinfelder, 2011) and indeed, coccolithophores may be limited by CO_2 under certain conditions (Ratti et al., 2007; Riebesell et al., 2007; Reinfelder, 2011). CA activity has been measured in coccolithophores (Sikes and Wheeler, 1982; Rost et al., 2003) although its involvement in a CCM is still unclear. It appears that *E. huxleyi* acquires HCO_3^- mainly by an anion exchange protein, but external CA can also be activated at low concentrations of DIC (Herfort et al., 2002).

The net production of CO_2 by coccolithophores is a nonlinear function of CO_2 liberation during $CaCO_3$ precipitation and carbon fixation *via* photosynthesis, and coccolithophores are likely neutral or represent a net sink for CO_2, except perhaps at very low concentrations of CO_2 (Reinfelder, 2011). The extent to which a calcifying organism represents a source or a sink of CO_2 to the environment is critically dependent on the production ratios of particulate inorganic carbon (PIC, often used interchangeably for $CaCO_3$) to particulate organic carbon (POC). Specifically, *E. huxleyi* represents a sink of CO_2 to the environment if the PIC:POC ratio is <1.5, but it is a source of CO_2 if the ratio is >1.5 (Frankignoulle et al., 1994). This ratio in *E. huxleyi* has been found to vary from 0.2 to >1 (Riebesell et al., 2000; Zondervan et al., 2001, 2002; Engel et al., 2005; Feng et al., 2008; Iglesias-Rodriguez et al., 2008), indicating that *E. huxleyi* likely acts as a CO_2 reservoir.

Calcification/photosynthesis interactions

Several studies have investigated the possible physiological coupling between photosynthesis and calcification, although these feedbacks are still unclear. The generation of CO_2 through calcification has been proposed as a possible CCM to increase the CO_2 availability at the site of RuBisCO, not only in coccolithophores (Buitenhuis et al., 1999) but also in other calcifying organisms including calcareous algae (Ries, 2010). In fact, although some calcifiers can accumulate sufficient DIC for photosynthesis in the absence of calcification, it has been suggested that, under nutrient- and light-replete conditions, calcification may partly support photosynthesis (Reinfelder et al., 2011). A general consensus has, however, not been reached and other studies have found that photosynthesis and calcification are decoupled in coccolithophores (Buitenhuis et al., 2001) and corals (Gattuso et al., 2000). Additionally, whilst calcification in *E. huxleyi* has been found to occur predominantly during the light phase (Paasche, 1962, 1964, 1965, 1966; Sikes et al., 1980; Linschooten et al., 1991), in some cases it can continue in the dark (Trimborn et al., 2007; Leonardos et al., 2009). Furthermore, in experiments using media lacking Ca^{2+} to prevent $CaCO_3$ precipitation, no effect was detected on organic carbon production suggesting that photosynthesis can be decoupled from calcification (Herfort et al., 2002.,Trimborn et al., 2007; Leonardos et al., 2009).

Genomics of calcification

Almost nothing was known about the molecular biology underpinning the tightly controlled intracellular process of calcification when Paasche synthesized his review in 2002. The only potential protein identified was GPA, a highly acidic Ca^{2+} binding molecule named for its high concentration of glutamic acid, proline and alanine residues (Cortstjens et al., 1998). However, the role of GPA is still putative and recent studies find conflicting results on its expression in relation to calcification (Quinn et al., 2006; Richier et al., 2009; von Dassow et al., 2009; Mackinder et al., 2011; Rokitta et al., 2011; Benner et al., 2013; Kegel et al., 2013).

Since 2002, a variety of genomic studies have attempted to look for proteins involved in biomineralization. Initial efforts involved growing cells under nitrate or phosphate limitation to stimulate calcification (see section "Macronutrient effects on physiology and ecology: phosphorus and nitrogen") and identify differentially expressed transcripts using methods such as EST analysis (Wahlund et al., 2004), suppressive subtractive hybridization (Nguyen et al., 2005), SAGE (Dyhrman et al., 2006) and microarrays (Quinn et al., 2006). These studies showed that proteins such as CA (Quinn et al., 2006) and calcium binding proteins (calnexin and calretulin, Wahlund et al., 2004) had potential roles in calcification, although results were difficult to interpret because of possible interactions with nitrate and phosphate starvation. More recent attempts have applied single or high-throughput transcript-based techniques to the haploid and diploid phases of *E. huxleyi* (Richier et al., 2009; van Dassow et al., 2009; Mackinder et al., 2011; Rokitta et al., 2011) assisted by the genome sequencing of strain CCMP1516 (Read et al., 2013), which enabled more extensive transcript annotation. Since only diploid cells calcify, the premise was that highly up-regulated transcripts from this phase could be involved in calcification. These studies have attempted to shed light on transporters that may have roles in providing Ca^{2+} and carbon substrate biomineralization (reviewed by MacKinder et al., 2010).

Despite best molecular efforts, still little is known about cellular Ca^{2+} transport dynamics in *E. huxleyi* and EST-based methods have failed to find any Ca^{2+}-ATPases associated with calcified diploid cells (von Dassow et al., 2009). However, transcripts from a Ca^{2+}/H^+ VCX1-type antiporter and NCKX proteins associated with rapid Ca^{2+} efflux (von Dassow et al., 2009), as well as CAX3 (a Ca^{2+}/H^+ transporter, von Dassow et al., 2009; Mackinder et al., 2011; Rokitta et al., 2011) were found in diploid cells. The latter has been proposed to have a role during transport of Ca^{2+} to the endomembrane system for calcification (Mackinder et al., 2011). The up-regulation of solute carrier 4 (SLC4) in diploid, calcified cells (also known as AEL1, a Cl^-/HCO_3^- transporter, von Dassow et al., 2009; Mackinder et al., 2011; Rokitta et al., 2011) suggested that this protein is involved in calcification related HCO_3^- dynamics, either enabling HCO_3^- to cross the plasma membrane or enter cellular compartments involved in biomineralization (von Dassow et al., 2009; Mackinder et al., 2011). Interestingly, RNA-seq analysis showed that none of these genes were differentially expressed in *E. huxleyi* cells presenting higher calcification rates under different CO_2 levels in long-term chemostats (>700 generations, Benner et al., 2013) since both CAX3 and NCKX2 were expressed at high levels in all treatments. The lack of a confirmed trend for carbon substrate or Ca^{2+} uptake means that it is plausible that many transporters involved in calcification are as yet undiscovered.

Cytological dynamics during biomineralization also remain relatively unknown for *E. huxleyi*. On the basis of proteomic and transcriptomic data, clathrin-mediated endocystosis/vesicular trafficking has also been proposed to be involved in calcification (Jones et al., 2011; Rokitta et al., 2011), whereas Quinn et al. (2006) suggested that α-soluble *N*-ethylmalemimide sensitive factor attachment (SNAP) proteins and RabGAP/TBC-containing proteins may have a role in SNARE-type secretory dynamics during coccolithogenesis. Additionally, von Dassow et al. (2009) and Rokitta et al. (2011) found SNARE-type homologs associated with calcifying diploid cells. Haploid/diploid studies indicated up-regulation of components of vacuolar type H^+-ATPases in calcified diploid cells (von Dassow et al., 2009; Mackinder et al., 2011; Rokitta et al., 2011), and comparative genome hybridization (CGH) of sixteen strains recently indicated that that up-regulation of two V-type protein ATPase proteolipid subunits was associated with calcification status (Kegel et al., 2013). Activity of these pumps was previously located to *Pleurochyrsis* coccolith vesicles (Araki and Gonzalez, 1998) and subsequently subunits were immunolocalized to this cellular component (Cortsjens et al., 2001). It has been proposed that V-ATPases are involved in alkalinizing vesicles to enable a suitable pH for coccolithogenesis but recent RNA-seq investigations indicated that these components were not always associated with calcification (Benner et al., 2013). Recently, Mackinder et al. (2010) challenged the theory of alkalization by proposing that V-ATPases were involved in coccolith vesicle acidification, with an influx of H^+ enabling the transport of Ca^{2+} to the site of calcification through CAX3 transporters, which was dismissed by Holtz et al. (2013; see section "Inorganic carbon and carbon kinetics"). It is clear the genomics unpinning substrate transport and pH regulation in *E. huxleyi* require much more investigation.

Whilst most researchers have focused their attention on transporters, components associated with $CaCO_3$ during biomineralization still remain understudied in *E. huxleyi*. It is likely that coccolith morphogenensis is controlled in part by the cytoskeleton, since the inhibition of microtubules and actin results in the formation of malformed coccoliths (Langer et al., 2010). The identification of proteins with similarities to those involved in biomineralization within other organisms (e.g. Nguyen et al., 2005) indicates that many recent studies still contain a wealth of information waiting to be gleaned. However, time and time again, exploratory molecular studies uncover a substantial number of unknown genes and proteins containing sequences that cannot be found in any current genomic database or unusual proteins with conserved motifs (e.g. Wahlund et al., 2004; Nguyen et al., 2005; Quinn et al., 2006; Benner et al., 2013; Jones et al., 2013). Data from these studies may be a useful starting point for future investigations into the molecular dynamics of calcification; for example it is possible that genes containing a Ca^{2+} binding domain and a Ca^{2+}/Mg^{2+} channel, exhibiting homology to the connective tissue glycoprotein fibrillin, are involved in coccolith biomineralization (Benner et al., 2013). Additionally, proteins involved in vitamin K2 biosynthesis were found within the *E. huxleyi* proteome (Jones et al., 2013), a promising target since vitamin K2 plays a role in the process of Ca^{2+} chelation and biomineralization in human osteoblasts (Akedo et al., 1992). It has also been suggested that chitinase and chinodextrinase have a role in the dynamics of chitin-like polysaccharides that could bind coccoliths to the organic extracellular surface of *E. huxleyi* cells (Benner et al., 2013). Additionally, CGH analysis indicated that a long chain fatty acid CoA ligase, an arylsulfatase with a Ca^{2+} binding subunit, an uncharacterized oxidoreducatase, an activator of a 90 kDa heat shock protein ATPase and a kelch-like protein were associated with calcification (Kegel et al., 2013). It is plausible that some or all of these proteins could have roles in the process of biomineralization and these studies remain a wealth of information for gene hunting. However, in the absence of a viable transformation mechanism for *E. huxleyi*, function and confirmation of their role may remain speculative.

Ecophysiology and responses to environmental change

Since Paasche (2002), a substantial body of literature has appeared on the effect of environmental change and future ocean conditions on *E. huxleyi* growth and calcification. Specific responses in relation to temperature, salinity, irradiance, nutrient availability and carbon chemistry will be reviewed in this section. Recent molecular studies regarding irradiance will be briefly described and for a thorough overview of the response of this species to light, particularly with respect to calcification, we direct the reader to Zondervan (2007). As previously stressed, general assumptions on physiology should be made with caution considering the vast levels of diversity that exist at the molecular level between strains (see section "Genetic and morphological diversity"). Additionally in the years since the Paasche review, multi-parametric studies are becoming more common in the literature, yielding a more complete picture of how environmental dynamics affect *E. huxleyi* physiology.

Temperature

Emiliania huxleyi has been observed globally from subtropical to subpolar waters. Recent observations indicate that the species may also be present at low densities in Southern Ocean waters below 60°S (e.g. Cubillos et al., 2007; Gravalosa et al., 2008) and in high Arctic waters (80°N; Heimdel, 1983; Hegseth and Sundfjord, 1998; Charalampopoulou et al., 2011). Satellite observations also indicate a poleward expansion of *E. huxleyi* between 1997-2007 (Winter et al., 2014). These findings challenge previous paradigms regarding the distribution of this species that were described by Paasche in 2002 but the extent to which this relates to climate change, physiology, current dynamics or is the result of higher resolution sampling remains difficult to determine. However, the near uniform geographical distribution of this species suggests that local populations exist adapted to different hydrographic regimes.

Temperature has been considered as a primary control factor setting the boundaries of *E. huxleyi* biome (Boyd et al., 2010) and the distribution of morphotypes (Mohan et al., 2008; Henderiks et al. 2012; see section "Morphotype diversity"), and many studies have investigated how regional projections of warming might alter *E. huxleyi* physiological performance. As reviewed by Fielding (2013), the specific growth rate of *E. huxleyi* generally increases between 2°C and 27°C, with maximum growth rates equaling ~2 d^{-1}, although data are typically clone specific (Paasche, 2002). Growth data do not exist for temperatures below 2°C and the few studies conducted above 27°C have showed growth rates close to zero (Fielding, 2013). Although most models on *E. huxleyi* bloom formation use an exponential relationship to define the relationship between the maximum growth rate (μ_{max}, d^{-1}) and temperature (*T*, °C), a meta-analysis indicated that a power function ($\mu_{max} = 0.1419T^{0.8151}$) seemed to be a more appropriate relationship to use in models incorporating temperature-dependent maximum growth rates of *E. huxleyi* (Fielding, 2013).

Emiliania huxleyi seems to show a variety of responses to temperature with respect to calcification, however no molecular studies to date have been conducted on the mechanisms behind this variability. A decrease in temperature from 20 to 10°C enhanced cell protoplast, coccosphere size and coccolith production for strain EH2 (Sorrosa et al., 2005), suggesting that calcification might increase as growth rate decreases. Conversely, an increase in temperature from 13 to 17°C decreased calcification rates and coccosphere size for strain AC481 with no change in coccolith size, suggesting a lower $CaCO_3$ content per coccolith, a decrease in coccolith number per cell and/or a decrease in coccolith production rate under high temperatures (de Bodt et al., 2010). However since an increase from 10 to 20°C did not affect coccolith size for strain PLY B92/11 (Fielding et al., 2009), it is important to recognize that strains differ in their thermal optimum and physiological responses. Also, most physiological studies have focused on coastal strains (Fielding et al., 2013), isolated from waters with a wide thermal window, and therefore might demonstrate an enhanced ability to adapt to increasing water temperatures (Huertas et al., 2011).

Salinity

Reflecting its global distribution pattern, *E. huxleyi* is found at salinities ranging from 11 to 41‰ (Winter et al., 1994). Although salinity does not appear to be an important factor defining bloom formation of this species, it seems to have an effect on the regulation of calcification, although the underlying mechanisms remain unclear. Recent morphometric analysis of globally distributed sediment core tops, plankton samples and culture experiments demonstrated a positive correlation between *E. huxleyi* coccolith size and sea surface salinities between 33 to 38‰ (Bollmann and Herrle, 2007; Bollmann et al., 2009; Fielding et al. 2009) suggesting that coccolith morphology could be used as a proxy for paleo-surface salinity reconstructions. However, this linear relationship has only been resolved for open-ocean data, and coccoliths from marginal environments (e.g. Norwegian coastal waters, Black Sea and Northern Red Sea) deviate from this correlation (Fielding et al 2009). This again reflects the possible variation in the adaptive capabilities of different *E. huxleyi* morphotypes and clones linked to biogeography and evolutionary history. In fact, as reviewed by Paasche (2002), laboratory studies revealed that coastal clones were generally less tolerant to salinity changes than coastal ones. Further analysis of *E. huxleyi* strains from both marginal and open-ocean settings are required to gain more insights into coccolith morphology and sea surface salinity relationships.

Irradiance

Growth rates of *E. huxleyi* are light dependent and typically increase as light level increases, until a certain limit when growth becomes saturated (see review by Zondervan, 2007). Light is an important requirement for *E. huxleyi* bloom formation and blooms are primarily confined to the top 10-20 m of the water column, associated with high light conditions (e.g. Nanninga and Tyrrell, 1996). In contrast to other phytoplankton groups, *E. huxleyi* does not show photoinhibition at irradiances up to 2 500 µmol photons m^2 s^{-1} (Zondervan, 2007), and saturation irradiances are higher than those for dinoflagellates or diatoms (as described by Paasche, 2002). Recent molecular studies have revealed that photoacclimation in *E. huxleyi* involves a trade-off between light absorption and photoprotective characteristics that might include changes in abundance and composition of proteins in the light-harvesting antenna of PSII (Lefebvre et al., 2010; McKew et al., 2013a, b). However, only ~20% of the proteome was differentially regulated when cells were grown at 30 and 1 000 µmol photons m^2 s^{-1} (McKew et al., 2013b) and in intermittent light, *E. huxleyi* seems to maintains a stable core of catalysts involved in CO_2 fixation, respiratory energy metabolism and biosynthesis. These differences confer a competitive advantage to *E. huxleyi* in stratified waters. Greater resilience to photoinhibition might also benefit *E. huxleyi* under future climate conditions, where mixed layer depths are expected to be shallower, causing phytoplankton to be closer to the surface and exposed to higher light irradiances.

Dissimilarities between *E. huxleyi* and diatom pigment composition may explain differences in light harvesting efficiency and photoprotective abilities between these two groups (van de Poll et al., 2007). Calcification may also provide an additional advantage for *E. huxleyi*; sudden increases in light intensity produce an excess of energy that might be used to channel CO_2 towards the proximities of RuBisCO, as well as Ca^{2+} and inorganic carbon into calcification vesicles. For example, an abrupt increase in light from 50 to 800 µmol.m^2.s^{-1} prompted a rapid (11-fold) increase in calcification rate by *E. huxleyi* in the first 10 minutes of light exposure, which was two-fold more than the increase in the carbon fixation rate during the same period (Barcelos e Ramos et al., 2012). In contrast, cells of the diatom *Phaeodactylum tricornutum* used the extra energy to increase carbon fixation rates (Barcelos e Ramos et al., 2012). Although the role of coccolithophore calcification is still uncertain, this result might suggest a potential function in dissipating excess energy during cellular acclimation to new light conditions. The presence of coccoliths has also been proposed to serve as a light scatter mechanism to reduce cellular photo-damage at high irradiances (Gao et al., 2009).

Intraspecific differences in high light responses once again emphasize the complexity present within the *E. huxleyi* species group. For example, as light levels increase, most phytoplankton groups reduce the number of photosynthetic units (PSU) present per cell whilst keeping PSU size constant (Suggett et al., 2007). However, an increase in light levels caused the calcifying strain B92 to preferentially decrease PSU size instead of number. Meanwhile the non-calcifying strain (B11) decreased both PSU size and number to maintain a relatively constant balance of absorption energy between photosystem II (PSII) and PSI under high light (Suggett et al., 2007). These responses are different to those exhibited by other microalgae and provide a basis to understanding the flexible acclimation mechanisms of *E. huxleyi* under high light. Additionally, the intraspecific variability in photoacclimation responses likely reflects genetic differences related to evolutionary history and calcification status.

Macronutrient effects on physiology and ecology: phosphorus and nitrogen
Emiliania huxleyi is known to flourish after the demise of the diatom bloom, when nitrate, phosphate and silicate are limiting, and populations are predominantly found in stratified surface waters under conditions of moderate to low nutrient availabilities (Egge and Heimdal, 1994; Marañon et al., 1996; Litchman et al., 2006; Siegel et al., 2007; Leblanc et al., 2009). As reviewed by Paasche (2002), the ecological advantage that *E. huxleyi* possesses in oligotrophic waters may stem from low nutrient quotas, an extremely high affinity for orthophosphate under phosphate-limiting conditions, and the expression of a constitutive an inductive alkaline phosphatase, allowing for the acquisition of inorganic phosphorus from organic phosphorus pools. *E. huxleyi* is also able to better maintain PSII function under nitrogen depletion compared to diatoms (Loebl et al., 2010). The species is typically a poor competitor for ammonium and nitrate, however some strains are able to use a variety of dissolved organic nitrogen (DON) compounds such as amides (Palenik and Henson, 1997), or other organic nitrogen and phosphorus compounds (e.g. glycine, urea, glycerophosphate) that other coccolithophore species, such as *Coccolithus braarudii* and *Calcidiscus leptoporus*, cannot use (Benner et al., 2010).

Since Paasche (2002), both physiological and molecular studies have improved the knowledge about nitrogen and phosphorus preferences and regulation of macronutrient metabolism in *E. huxleyi*. Along with an increase in alkaline phosphate activity, three phosphate-regulated proteins (PRP1, PRP2 and PRP3) were detected by the biotinylation of cell-surface proteins in *E. huxleyi* under phosphate limiting conditions (Dyhrman and Palenik, 2003), and it was suggested that PRP3 might possess 5′-nucleotidase activity. Intraspecific differences in alkaline phosphatase activity and protein sequence may result in physiological differences between different strains under low phosphate conditions (Dyhrman and Palenik, 2003; Xu et al., 2006; Landry et al., 2009). SAGE analysis further identified potential genes and molecular pathways involved in nutrient scavenging by *E. huxleyi* and highlighted the potential importance of polyphosphate as a short-term storage pool for phosphorus (Dyhrman et al., 2006). Recent physiological studies have also revealed interesting mechanistic effects of phosphorus limitation. For example, under phosphorus scarcity, *E. huxleyi* was able to interchange phosphorus lipids for those that do not contain phosphorus, such as betaine lipids, a mechanism that reduces total cellular phosphorus demand by up to 30% (Van Mooy et al., 2009).

Although phosphorus metabolism has been fairly extensively studied in *E. huxleyi*, less is known about enzymes or transporters involved in nitrogen metabolism and its regulation. Iwamoto and Shiraiwa (2003) characterized an *E. huxleyi* nitrate reductase (NR), which exhibited novelty in its subunit structure and kinetic properties when compared to NR sequences from other algae and plants. Landry et al. (2009) investigated the up-regulation of a putative amidase located on the cell surface, which appeared to be involved in the acquisition of nitrogen from amino acids under nitrate limiting conditions. This protein was expressed at the end of an *E. huxleyi* bloom in Norway and was suggested as an indicator of the nitrogen status in this species (Landry et al., 2009).

Recently, Bruhn et al. (2010) investigated the expression of a putative formamidase (FMD), a glutamine synthetase (GS) and an NR when cells of *E. huxleyi* were grown under different nitrogen sources. This indicated that *E. huxleyi* uses a transcript-related nitrogen preference hierarchy. When nitrate is abundant, low priority is given to the catabolism of organic nitrate into ammonium by FMD or the assimilation of nitrate into glutamine by GS (Bruhn et al., 2010). Recently, Kaffes et al. (2010) observed than when *E. huxleyi* was grown under concentrations of nitrate typical of the North Atlantic Ocean during spring (~10 µM), cells adjusted the kinetic parameters of NR, nitrite reductase, GS and glutamate synthase to coordinate intracellular nitrate and CO_2 fluxes and thus maintain constant intracellular nitrogen and carbon ratios.

Interactions between calcification and ambient macronutrient concentration have been repeatedly observed in *E. huxleyi* and it is thought that calcification is enhanced by nutrient limitation. However, the mechanisms underlying the interplay between nutrient status and the transformation of inorganic carbon into $CaCO_3$ remain largely unknown. Paasche (2002) and Shiraiwa (2003) suggested that *E. huxleyi* blooms in the ocean might be triggered by a supply of nutrients that stimulate organic matter production and cell growth, later leading to nutrient depletion suppressing growth and enhancing calcification. Müller et al. (2008) confirmed that calcification in *E. huxleyi* is tightly linked to the G1 phase of the cell cycle. During this phase, cells have a high nitrogen requirement in order to increase biomass and reach the size threshold for division. Consequently, nitrogen limitation results in a reduced cell size and a tendency for cells to remain in the G1 phase, resulting in an increase in cellular $CaCO_3$ content. However, phosphorus limited cells were larger and more calcified since biomass buildup continued, but DNA synthesis and replication were likely inhibited, preventing cells from entering the S phase (Müller et al. 2008).

It is important to clarify that although $CaCO_3$ and POC cellular quotas vary under conditions of nutrient limitation, $CaCO_3$ and POC production might follow different patterns. Also, considering that the distribution of *E. huxleyi* morphotypes might be partly dependent on ambient oceanic nutrient concentrations (Mohan et al., 2008; Henderiks et al., 2013), the use of different morphotypes should be considered to gain a broader perspective.

Effect of inorganic carbon chemistry on Emiliania huxleyi physiology

Over the period known as the Anthropocene (Crutzen and Stoermer, 2000), CO_2 of anthropogenic origin has been rising in the atmosphere and the concentration of CO_2 in seawater has increased proportionally as it has diffused into the ocean. Some of this dissolved CO_2 reacts with water to form carbonic acid (H_2CO_3), a process termed ocean acidification (Caldeira and Wickett, 2003, 2005). These chemical shifts have been found to have an effect on the acclimation of *E. huxleyi* such as modulation of its physiology (e.g., changes in photosynthesis, growth rates and calcification) in response to ocean acidification (e.g., Langer et al., 2009) and selection of morphotypes with distinct inherent calcification properties (e.g., Beaufort et al., 2011; Bach et al., 2012; Henderiks et al., 2012). Given the geographical extent and abundance of *E. huxleyi*, these shifts in cell physiology also have the potential to alter marine biogeochemical cycles.

Over the last decade, since the term ocean acidification was coined (Caldeira and Wickett, 2003), there has been a prolific generation of manuscripts dealing with the reciprocal interactions between the biology of coccolithophores and the changing inorganic carbon chemistry in the oceans. Much of this work concerns the regulation of calcification and stems from our poor understanding of its biological and evolutionary roles. Indeed, this metabolic function is still under scrutiny in the fields of biology, geochemistry and genomics. Negative effects of ocean acidification on coccolithophores have been observed in lab and field studies but the responses appear to be species-specific and in some cases strain-specific (e.g., Riebesell et al., 2000; Iglesias-Rodriguez et al., 2008; Grelaud et al., 2009; Langer et al., 2009; de Bodt et al., 2010; Beaufort et al., 2011). Meta-analyses have confirmed a decline in calcification with exacerbated ocean acidification but also revealed a great degree of sensitivities to increasing seawater CO_2 levels and high diversity of responses in all calcifying groups including coccolithophores (Kroeker et al., 2010).

Among the main concerns about the future of coccolithophores and other calcifiers is the extent to which the recent exceptionally fast alterations in carbon chemistry, compared to the rate of alteration over the last 65 million years (Ridgwell and Schmidt, 2010), will affect the function and resilience of coccolithophores. Whereas the chemistry of ocean acidification is relatively well understood, much less is known about the biological and adaptive effects of these chemical changes. To date, the main gaps in our knowledge stem from the poorly known molecular/biochemical mechanisms of biomineralization and the challenges in deciphering biological $CaCO_3$ precipitation from abiotic dissolution driven by ocean acidification (Iglesias-Rodriguez, 2013).

It has been argued that the shifts in carbonate chemistry associated with ocean acidification often demand energetically costly metabolic adjustments (e.g., Irie et al., 2010). Genome sequencing and long term laboratory manipulations are beginning to shed light into the extent to which *E. huxleyi* populations adapt over time. Information at all levels of population organization, from genes to phenes at varying time scales, will enhance our ability to predict the fate of *E. huxleyi* in a high CO_2 environment.

Future directions

As we have discovered, discrepancies in calcification responses to simulated ocean acidification are probably just one manifestation of the intraspecific differences between *E. huxleyi* strains. As we move forward to think about how best to understand the physiological ecology of this important organism, we must also take into account that phenotypic and genomic properties of strains are known to vary over time under continuous culturing (Lakeman et al., 2009), which could cause identical strains to diverge between laboratories. Recently, some laboratories have addressed issues of phenotypic drift in long term experiments and have ensured that more realistic multi-parametric experiments are conducted to better help us understand responses of this important species to environmental perturbations.

Short-term versus long-term experiments

Emiliania huxleyi, along with many other phytoplankton species, has a short generation time and large population size, conferring the species with a high potential for acclimation and adaptation to environmental change (Elena and Lenski, 2003). Responses from short-term studies might, therefore, vary from those of long-term experiments since the latter allow for evolutionary responses and the selection of new clones or genotypes that might randomly arise as result of spontaneous mutation (Lohbeck et al., 2012). To date, most studies have investigated the response of *E. huxleyi* to different environmental parameters in short-term experiments. For example, only a few studies have tested the long-term responses of *E. huxleyi* to the single effect of CO_2 (>150 generations, Müller et al. 2010; >500 generations, Lohbeck et al, 2012), the combined effect of CO_2 and nitrogen source (>200 generations, Lefebvre et al. 2012), or CO_2 and temperature (>700 generations, Benner et al. 2013). Both Müller et al. (2010) and Lohbeck et al. (2012) found a similar reduction in growth and calcification rate after short and long-term incubations. However, growth rates of cells grown under high CO_2 for 500 generations were higher than those of cells maintained at current-day CO_2 when both populations were briefly exposed to high CO_2, indicating possible selection for genotypes with better fitness under high CO_2 (Lohbeck et al., 2012). Similarly, Benner et al. (2013) found that CCMP371 exhibited an increased inorganic carbon content and calcification rate under warm and acidified conditions after >700 generations, opposite to the effects of short-term exposure (Feng et al., 2008). These studies suggest that a deep-time perspective is needed to understand long-term effects of high CO_2 and help us estimate the speed and capacity of adaptation of *E. huxleyi.*

Multiparametric experiments

Laboratory studies assessing the effect of independent environmental parameters on the physiology of *E. huxleyi* are central to understanding the cellular mechanisms underlying physiological changes. However, reciprocal interactions between environmental parameters can yield different effects than individual variables alone, which can also limit the comparability of studies. When Paasche (2002) published his review, most studies had a uni-parametrical focus. However, there has been an increasing trend for multi-parametrical studies in recent years, especially those combining increasing pCO_2 with another variable (e.g. light, temperature, phosphorus or nitrogen availability). For instance, light may modulate the physiological effects of increasing CO_2 on growth and calcification of *E. huxleyi* (see Zondervan, 2007). Cellular calcification of strain CCMP371 was reduced under elevated CO_2 conditions only when irradiance was saturating (Feng et al., 2008). In contrast, a more pronounced decrease in $CaCO_3$ production was observed under low light compared to high light levels for the diploid strain RCC1217 when exposed to enhanced CO_2 concentrations (Rokitta and Rost, 2012). Studies have also focused on the combined effects of increased CO_2 levels and temperature. For instance, the sole increase in CO_2 had more of an adverse impact (alteration of coccolith morphology) on strain AC481 than just the increase of temperature or the combination of both factors (de Bodt et al., 2010). In contrast, a combined increase in elevated CO_2 or temperature drastically decreased $CaCO_3$ production and enhanced *E. huxleyi* abundance in a natural North Atlantic Spring bloom, response that was not observed when these parameters were independently manipulated (Feng et al., 2009).

The response of *E. huxleyi* to elevated CO_2 has been intensely studied in laboratory experiments under nutrient replete conditions (see section "Effect of inorganic chemistry on *Emiliania huxleyi* physiology"). However, despite the importance of nutrients on *E.*

huxleyi growth and calcification and the projected extent of oligotrophic waters in future oceans, CO_2-related studies with limiting nutrients such as phosphorus and nitrogen have not been considered until recently (Sciandra et al., 2003; Leonardos and Geider, 2005; Borchard et al., 2011; Borchard and Engel, 2012; Lefebvre et al., 2012; Müller et al., 2012; Rouco et al., 2013). Coccolith production in these studies presented a different response to increasing CO_2 depending on the nutrient condition and the strain, although direct comparisons between strains are difficult to make since each study used different experimental approaches. In these studies, variations in growth and photosynthetic rate, coccolith morphology, organic carbon assimilation as well as cellular nutrient stoichiometry varied between treatments and strains. These results are important since growth rates and the biochemical composition of phytoplankton influence grazing selection (Jones et al., 2002), and this has the potential to determine the prevalence of some strains *versus* others in future oceans. Recent studies also assessed the interactive effect of the nitrogen source and enhanced CO_2 on carbon allocation, growth and morphology of CCMP371 (Lefebvre et al., 2012), observing that these two factors have a synergistic effect on *E. huxleyi* carbon partitioning. Studies such as this are necessary for an understanding of the response of *E. huxleyi* to future ocean conditions and for the parameterization of global biogeochemical models.

Conclusions and thoughts on how to move forward

In our updated review on the biology of *E. huxleyi*, we find that the community is at a crossroads regarding how to approach the vast levels of diversity that are beginning to be unveiled as modern sequencing technologies become widely embraced. We, along with other authors, find that a drastic re-evaluation of the species concept is required in marine phytoplankton biology. Genomes of strain EH2 and CCMP1516 differ by 30.1%, whereas the genera *Gephyrocapsa* and *Emiliania* differ by only 30.4% (Kegel et al., 2013). Can we associate these differences with biogeochemistry? Challenges facing the next generation of coccolithophore biologists involve understanding whether this diversity matters and how it can be best represented in models. For example, *E. huxleyi* blooms occur under widely varying environmental conditions suggesting either that different environmental factors trigger their periodic dominance, or that blooms in different environmental regimes represent periodic dominance of individual species within the diverse *E. huxleyi* phylogenetic group. Unraveling the nature of these cause-and-effect dynamics will likely require synthesis of genetic, biochemical, and physiological information in future years. The power of sequencing approaches in the post-genomic era makes collection of some of these data relatively straightforward, given proper experimental design. However, the anticipated promise of the sequencing revolution to resolve questions about algal ecophysiology is now recognized to depend on a more integrated understanding of genomics and physiological responses to environmental conditions.

As discussed, the recent genome comparison of 14 *E. huxleyi* isolates demonstrated that this species is far more diverse than previously thought (Kegel et al., 2013, Read et al., 2013). This is a discovery that has become a refrain of sorts in the field of microbial diversity. All phototrophs share basic metabolic units: including light harvesting pigments, components of the photosynthetic and respiratory electron transport chains, carbon fixation and catabolic pathways. Regulation of these metabolic units and their biochemical details can vary among algal groups and these differences are expected to lead to ecological niche differentiation. The genome sequence serves as a blueprint for the presence of the fundamental phototrophic metabolic units as well as unique features that underlie strain-specific biogeochemical roles. The goal is to identify these features, understand their environmental controls, and quantify their impact on growth. Although straightforward in principle, identification of these genomic features has proven challenging. In the case of *E. huxleyi*, 70% of genes in the core and variable genomes are of unknown function (Read et al., 2013). Furthermore, the biochemistry needed to confirm ambiguous annotations is a known bottleneck in functional genomics.

There is a huge gap between the knowledge of a phototroph's genome sequence and predictions of carbon accumulation rates (primary production) under varying environmental conditions. Whole cell gene and protein analyses (e.g., transcriptomics and proteomics, respectively) can provide views of subcellular activity that bridge the gap between the genome sequence and expressions of cell growth. These analyses are crucial to establishing molecular linkages between the genome and metabolism. They can also reveal key regulatory switches that operate under different growth conditions and be used to identify molecular proxies for estimating process rates in the field. Further complicating interpretations of next-generation sequence data are reports indicating that the transcriptome and proteome are generally decoupled (reviewed in Nie et al., 2007), reflecting regulatory processes that act at the levels of both transcription and translation. Indeed, recent CGH analysis of 16 strains of *E. huxleyi* showed that only 11 genes were associated with calcification (Kegel et al., 2013) indicating the potential importance of post-transcriptional control on this biogeochemically important function.

The field of systems biology applies computational approaches to understanding how cells utilize the complete collection of metabolic pathways leading to cellular growth. Some of these computational models integrate genome sequence information and expressions of cell growth (typically pigment content and/or elemental stoichiometry) with the goal of calculating the optimal growth rate under varying growth conditions. Other models include trade-offs that may allow for cell persistence over increased growth rate (Geider et al. 2009). Such models of cell growth become more robust through the addition of constraints determined through next generation sequencing technologies as well as by including multiple measurements of cell physiology (Figure 5.2). These constraints serve to limit the range of solutions available to optimize photosynthate distribution through the metabolic networks and maximize growth rate (Follows and Dutkiewics, 2012). Flux-based analysis has emerged as a powerful computational tool that is built upon the foundation of a complete genomic sequence and can be expanded through additional layers of genetic expression of increasing complexity (e.g., transcriptomics and proteomics; Price et al. 2003). Such detailed computational approaches are likely needed to fully understand the processes underlying *E. huxleyi* growth dynamics, and they will undoubtedly reveal unexpected strategies for metabolic adaptation under certain growth conditions (e.g., Dyhrman et al., 2012).

Even more promising for understanding biogeochemical impacts of *E. huxleyi* is the opportunity to connect genomics and physiology to expressions of growth that are retrievable from space. We envision that these scales of study can be bridged by

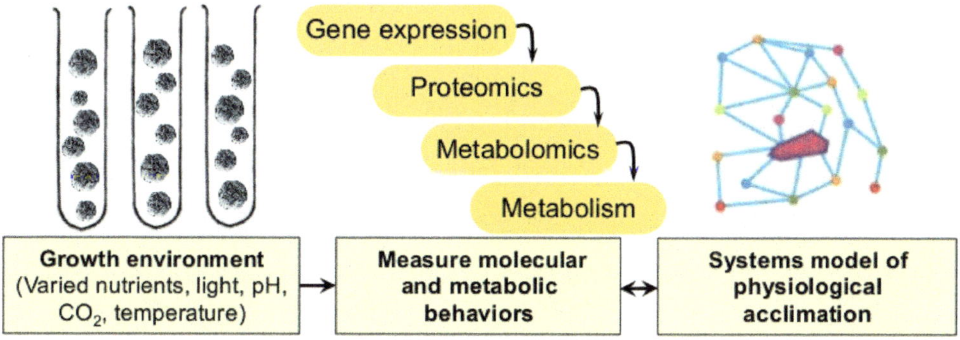

Figure 5.2. A systems biology approach to studying coccolithophore physiology. Controlled laboratory experiments to study phenotypic plasticity combined with genome and post-genomic sequence information provide the inputs and constraints for reconstructing metabolic networks. Computationally driven modeling is used to develop new hypotheses about metabolic mechanisms and predict growth responses to changing environments.

identifying key proxies of cellular status that can be linked to optical properties obtainable through remote sensing technologies. Thus, efforts to unravel the molecular biology controlling photosynthetic rates and primary production at the cellular level could merge with improved image analysis and interpretation of processes occurring at regional and global scales to yield meaningful understanding of *E. huxleyi* biogeography in an integrated, global context. We have come a long way since 2002 and as the community continues to uncover the secrets of this cosmopolitan species, we look forward to the next decade of *E. huxleyi* research.

Acknowledgements
We thank Jeremy Young and Alex Poulton for images, helpful suggestions and comments. We also thank Kristen Reifel for thoughts on an early draft of this chapter.

References

Akedo Y., Hosoi T., Inoue S., Ikegami A., Mizuno Y., Kaneki, M., Nakamura, T., Ouchi, Y., and Orimo, H. (1992). Vitamin K2 modulates proliferation and function of osteoblastic cells in vitro. Biochem. Biophys. Res. Commun. *187*, 814–820.

Araki, Y., and González, E.L. (1998). V- and P-type Ca^{2+}-stimulated ATPases in a calcifying strain of Pleurochrysis sp. (Haptophyceae). J. Phycol. *34*, 79–88.

Bach, L.T., Bauke, C., Meier, K.J.S., Riebesell, U., and Schulz, K.G. (2012). Influence of changing carbonate chemistry on morphology and weight of coccoliths formed by *Emiliania huxleyi*. Biogeosciences *9*, 3449–3463

Badger, M.R., Andrews, T.J., Whitney, S.M., Ludwig, M., Yellowlees, D.C., Leggat, W., and Price, G.D. (1998). The diversity and co-evolution of RuBisCO, plastids, pyrenoids and chloroplast-based CCMs in the algae. Can. J. Bot. *76*, 1052-1071.

Barcelos e Ramos, J., Schulz, K.G., Febiri, S., and Riebesell, U. (2012). Photoacclimation to abrupt changes in light intensity by *Phaeodactylum tricornutum* and *Emiliania huxleyi*: the role of calcification. Mar. Ecol. Prog. Ser. 452, 11-26.

Beaufort, L., Probert, I., and Buchet, N. (2007). Effects of acidification and primary production on coccolith weight: Implications for carbonate transfer from the surface to the deep ocean. Geochem. Geophys. Geosyst. *8*, Q08011.

Beaufort, L., Probert, I., de Garidel-Thoron, T., Bendif, E.M., Ruiz-Pino, D., Metzl, N., Goyet, C., Buchet, N., Coupel, P., Grelaud, M., Rost, B., Rickaby, R.E.M., and de Vargas, C. (2011). Sensitivity of coccolithophores to carbonate chemistry and ocean acidification. Nature *476*, 80–83.

Benner, I., and Passow, U. (2010). Utilization of organic nutrients by coccolithophores. Mar. Ecol. Prog. Ser. *404*, 21-29.

Benner, I., Diner, R.E., Lefebvre, S.C., Li, D., Komada, T., Carpenter, E.J., and Stillman, J.H. (2013). *Emiliania huxleyi* increases calcification but not expression of calcification-related genes in long-term exposure to elevated temperature and pCO_2. Phil. Trans. R. Soc. B. *368*, 20130049.

Billard, C., and Inouye, I. (2004). What is new in coccolithophore biology? In Coccolithophores - From Molecular Processes to Global Impact, H.R. Thierstein and J.R. Young, eds. (New York, U.S.A.: Springer), pp. 1- 30.

Bollmann, J., and Herrle, J.O. (2007). Morphological variation of *Emiliania huxleyi* and sea surface salinity, Earth Planet. Sci. Lett. *255*, 273–288.

Bollmann, J., Herrle, J.O., Cortés, M.Y., and Fielding, S.R. (2009). The effect of sea water salinity on the morphology of *Emiliania huxleyi* in plankton and sediment samples. Earth Planet. Sci. Lett. *284*, 320–328.

Borchard, C., Borges, A., Händel, N., and Engel, A. (2011). Biogeochemical response of *Emiliania huxleyi* (PML B92/11) to elevated CO_2 and temperature under phosphorus limitation: A chemostat study. J. Exp. Mar. Bio. Ecol. *410*, 61-71.

Borchard, C., and Engel, A. (2012). Organic matter exudation by *Emiliania huxleyi* under simulated future ocean conditions. Biogeosciences *9*, 3405-3423.

Boyd, P.W., Strzepek, R., Fu, F., and Hutchins, D.A. (2010). Environmental control of open ocean phytoplankton groups: Now and in the future. Limnol. Oceanogr. *55*, 1353–1376.

Brand, L.E. (1982). Genetic variability and spatial patterns of genetic differentiation in the reproductive rates of the marine coccolithophores *Emiliania huxleyi* and *Gephyrocapsa oceanica*. Limnol. Oceanogr. *27*, 236–245.

Brown, C.W., and Yoder, J.A. (1994). Coccolithophorid blooms in the global ocean. J. Geophys. Res. *99*, 7467-7482.

Bruhn, A., La Roche, J., and Richardson, K. (2010). *Emiliania huxleyi* (Prymnesiophyceae): nitrogen-metabolism genes and their expression in response to external nitrogen sources. J. Phycol. *46*, 266-277.

Buitenhuis, E.T., de Baar, H.J.W., and Veldhuis, M.J.W. (1999). Photosynthesis and calcification by *Emiliania huxleyi* (Prymnesiophyceae) as a function of inorganic carbon species. J. Phycol. *35*, 949-959.

Buitenhuis, E.T., van der Wal, P., and de Baar, H.J.W. (2001). Blooms of *Emiliania huxleyi* are sinks of atmospheric carbon dioxide: a field and mesocosm study derived simulation. Glob. Biogeochem. Cy. *15*, 577-587.

Caldeira, K., and Wickett, M.E. (2003). Anthropogenic carbon and ocean pH. Nature *425*: 365.

Caldeira, K., and Wickett, M.E. (2005). Ocean model predictions of chemistry changes from carbon dioxide emissions to the atmosphere and ocean. J. Geophys. Res. *110*, C09S04.

Campbell, L., Shapiro, L.P., and Haugen, E. (1994). Immunochemical characterization of eukaryotic ultraplankton from the Atlantic and Pacific Oceans. J. Plankton Res. *16,* 35–51.

Charalampopoulou A., Poulton A J., Tyrrell T., and Lucas M.I. (2011). Irradiance and pH affect coccolithophore community composition on a transect between the North Sea and the Arctic Ocean. Mar. Ecol. Prog. Ser. *431*, 25-43.

Cook, S.S., Whittock, L., Wright, S.W., and Hallegraeff, G.M. (2011). Photosynthetic pigment and genetic differences between two Southern Ocean morphotypes of *Emiliania huxleyi* (Haptophyta). J. Phycol. *47*, 615–626.

Cook, S.S., Jones, R.C., Vaillancourt, R.E., and Hallengraeff, G.M. (2013). Genetic differentiation among Australian and Southern Ocean populations of the ubiquitous coccolithophore *Emiliania huxleyi* (Haptophyta). Phycologia *52*, 368–374.

Cooper, T.G., Filmer, D., Wishnick, M., and Lane, M.D. (1969). The active species of "CO_2" utilized by ribulose diphosphate carboxylase. J. Biol. Chem. *244*, 1081–83.

Corstjens, P.L.A.M., van der Kooij, A., Linschooten, C., Brouwers, G.-J.,. Westbroek, P., and de Vrind-de Jong, E.W. (1998). GPA, a calcium-binding protein in the coccolithophorid *Emiliania huxleyi* (Prymnesiophyceae). J. Phycol. *34,* 622-630.

Corstjens, P.L.A.M., Araki, Y., and Gonzalez, E.L. (2001). A coccolithophorid calcifying vesicle with a vacuolar-type ATPase proton pump: cloning and immunolocalization of the Vo subunit *c*1. J. Phycol. *37*, 71-78.

Crutzen, P.J., and Stoermer, E.F. (2000). The 'Anthropocene'. Global Change Newsletter *41*, 17–18.

Cubillos, J., Wright, S., Nash G., de Salas, M., Griffiths, B., Tilbrook, B., Poisson, A., and Hallegraeff, G.M. (2007). Calcification morphotypes of the coccolithophorid *Emiliania huxleyi* in the Southern Ocean: changes in 2001 to 2006 compared to historical data. Mar. Ecol. Prog. Ser. *348*, 47−54.

De Bodt, C., Van Oostende, N., Harlay, J., Sabbe, K., and Chou, L. (2010). Individual and interacting effects of pCO_2 and temperature on *Emiliania huxleyi* calcification: study of the calcite production, the coccolith morphology and the coccosphere size. Biogeosciences *7*, 1401-1412.

Dixon, G.K., Brownlee, C., and Merrett, M.J. (1989). Measurement of internal pH in the coccolithophore *Emiliania huxleyi* using 2',7'-Bis-(2-carboxyethyl)-5 (and-6) carboxyfluorescein acetoxymethylester and digital imaging microscopy. Planta *178*, 443–449.

Dyhrman, S.T., and Palenik, B. (2003). Characterization of ectoenzyme activity and phosphate-regulated proteins in the coccolithophore *Emiliania huxleyi*. J. Plankton Res. *25*, 1215-1225.

Dyhrman, S.T., Haley, S.T., Birkeland, S.R., Wurch, L.L., Cipriano, M.J., and McArthur, A.G. (2006). Long serial analysis of gene expression for gene discovery and transcriptome profiling in the widespread marine coccolithophore *Emiliania huxleyi*. Appl. Environ. Microbiol. *72*, 252-260.

Dyhrman, S.T., Jenkins, B.D., Rynearson, T.A., Saito, M.A., Mercier, M.L., Alexander, H., Whitney, L., Drzewianowski, A., Bulygin, V.V., Bertrand, E.M., Wu, Z., Benitez-Nelson, C., and Heithoff, A. (2012). The transcriptome and proteome of the diatom *Thalassiosira pseudonana* reveals a diverse phosphorus stress response. PLOS ONE *7*, e33768.

Egge, J.K., and. Heimdal, B.R. (1994). Blooms of phytoplankton including *Emiliania huxleyi* (Haptophyta). Effects of nutrient supply in different N:P ratios. Sarsia *79,* 333-348.

Elena, S.F., and Lenski, R.E. (2003). Evolution experiments with microorganisms: the dynamics and genetic bases of adaptation. Nat. Rev. Genet. *4*, 457–69.

Engel, A., Thoms, S., Riebesell, U., Rochelle-Newall, E., and Zondervan, I. (2004). Polysaccharide aggregation as a potential sink of marine dissolved organic carbon. Nature *428*, 929–932.

Engel, A., Zondervan, I., Aerts, K., Beaufort, L., Benthien, A., Chou, L., Delille, B., Gattuso, J.-P., Harlay, J., Heemann, C., Hoffman, L., Jacquet, S., Nejgstgaard, J, Pizay, M.-D., Rochelle-Newall, E., Schneider, U., Terbrueggen, A., and Riebesell, U. (2005). Testing the direct effect of CO_2 concentration on a bloom of the coccolithophorid *Emiliania huxleyi* in mesocosm experiments. Limnol. Oceanogr. *50*, 493-507.

Feng, Y., Warner, M.E., Zhang, Y., Sun, Y., Fu, F.-X., Rose, J.M., and Hutchins, D.A. (2008). Interactive effects of increased pCO_2, temperature and irradiance on the marine coccolithophore *Emiliania huxleyi* (Prymnesiophyceae). Eur. J. Phycol. *43,* 87-98.

Feng, Y., Hare, C.E., Leblanc, K., Rose, J.M., Zhang, Y., DiTullio, G.R., Lee, P.A., Wilhelm, S.W., Rowe, J.M., Sun, J., Nemcek, M., Gueguen, C., Passow, U., Benner, I., and Hutchins, D. A. (2009). The effects of increased pCO_2 and temperature on the North Atlantic Spring Bloom: I. The phytoplankton community and biogeochemical response. Mar. Ecol. Prog. Ser. *388*, 13-25.

Fielding, S.R. (2013). *Emiliania huxleyi* specific growth rate dependence on temperature. Limnol. Oceanogr. *58*, 663–666.

Fielding, S.R., Herrle, J.O., Bollmann, J., Worden, R.H., and Montagnes, D.J.S. (2009). Assessing the applicability of *Emiliania huxleyi* coccolith morphology as a sea-surface salinity proxy. Limnol. Oceanogr. *54*, 1475-1480.

Findlay, C.S., and Girardeau, J. (2000). Extant calcareous nanoplankton in the Australian Sector of the Southern Ocean (austral summers 1994 and 1995). Mar. Micropaleontol. *40*, 417−439.

Fiorini, S., Middelburg, J.J., and Gattuso, J.-P. (2011). Testing the effects of elevated pCO_2 on coccolithophores (Prymnesiophyceae): comparison between haploid and diploid life stages. J. Phycol. *47,* 1281–1291.

Follows, M.J., and Dutkiewicz, S. (2011). Modeling diverse communities of marine microbes. Annu. Rev. Mar. Sci. *3*, 427-451.

Frada, M., Not, F., Probert, I., and de Vargas, C. (2006). $CaCO_3$ optical detection with fluorescent *in situ* hybridization: a new method to identify and quantify calcifying microorganisms from the oceans. J. Phycol. *42,* 1162–1169.

Frada, M., Probert, I., Allen, A.E., Wilson, W.H., and de Vargas, C. (2008). The 'Cheshire Cat' escape strategy of the coccolithophore *Emiliania huxleyi* in response to viral infection. Proc. Natl. Acad. Sci. U.S.A. *105*, 15944–15949.

Frada, M.J., Bidle, K.D., Probert I., and de Vargas, C. (2012). *In situ* survey of life cycle phases of the coccolithophore *Emiliania huxleyi* (Haptophyta). Environ. Microbiol. *14,* 1558–1569.

Frankignoulle, M., Canon, C., and Gattuso, J.-P. (1994). Marine calcification as a source of carbon dioxide: positive feedback of increasing atmospheric CO_2. Limnol. Oceanogr. *39*, 458-462.

Gattuso, J.-P., Reynaud-Vaganay, S., Furla, P., Romaine-Lioud, S., Jaubert, J., Bourge, I., and Frankignoulle, M. (2000). Calcification does not stimulate photosynthesis in the zooxanthellate scleractinian coral *Stylophora pistillata*. Limnol. Oceanogr. *45*, 246–250.

Gao, K., Ruan, Z., Villafañe, V.E., Gattuso, J.-P., and Helbling, E.W. (2009). Ocean acidification exacerbates the effect of UV radiation on the calcifying phytoplanker *Emiliania huxleyi*. Limnol. Oceanogr. *54*, 1855-1862.

Geider, R.J., Moore, C.M., and Ross, O.N. (2009). The role of cost-benefit analysis in models of phytoplankton growth and acclimation. *Plant Ecology and Diversity 2*, 165-178.

Geisen, M., Young, J.R., Probert, I., Sáez, A.G., Baumann, K-H., Bollmann, J., Cros, L., de Vargas, C., Medlin, L.K., and Sprengel, S. (2004). Species level variation in coccolithophores. In Coccolithophores - From Molecular Processes to Global Impact, H.R. Thierstein and J.R. Young, eds. (New York, U.S.A.: Springer), pp. 327–366.

Godoi, R.H.M., Aerts, K., Harlay, J., Kaegi, R., Ro, C.-U., Chou, L., and Van Grieken. R. (2009). Organic surface coating on coccolithophores - *Emiliania huxleyi*: Its determination and implication in the marine carbon cycle. Microchem. J. *91*, 266-271.

Gravalosa, J.M., Flores, J-A., Sierro, F.J. and Gersonde, R. (2008). Sea surface distribution of coccolithophores in the eastern Pacific sector of the Southern Ocean (Bellingshausen and Amundsen Seas) during the late austral summer of 2001. Mar. Micropaleontol. *69*, 16-25.

Green, J.C., Course, P.A., and Tarran, G.A. (1996). The life- cycle of *Emiliania huxleyi*: a brief review and a study of relative ploidy levels analysed by flow cytometry. J. Mar. Syst. *9,* 33–44.

Grelaud, M., Beaufort, L., Cuven, S., and Buchet, N. (2009). Glacial to interglacial primary production and El Nino-Southern Oscillation dynamics inferred from coccolithophores of the Santa Barbara Basin. Paleoceanography *24*, PA1203.

Hagino, K., Okada, H., and Matsuoka, H. (2005). Coccolithophore assemblages and morphotypes of *Emiliania huxleyi* in the boundary zone between the cold Oyashio and warm Kuroshio currents off the coast of Japan. Mar. Micropaleontol. *55*, 19–47.

Hagino, K., Mahdi Bendif, E., Young, J.R., Kogame, K., Probert, I., Takano, Y., Horiguchi, T., de Vargas, C., and Okada, H. (2011). New evidence for morphological and genetic variation in the cosmopolitan coccolithophore *Emiliania huxleyi* (Prymnesiophyceae) from the cox1b-ATP4 genes. J. Phycol. *47,* 1164-1176.

Hegseth, E.N., and Sundfjord, A. (2008). Intrusion and blooming of Atlantic phytoplankton species in the high Arctic. J. Mar. Syst. *74*, 108–119.

Heimdal, B.R. (1983). Phytoplankton and nutrients in the waters north-west of Spitsbergen in the autumn of 1979. J. Plankton Res. *5*, 901–918.

Henderiks, J., Winter, A., Elbraechter, M., Feistel, R., van der Plas, A., Nausch, G., and Barlow, R. (2012). Environmental controls on *Emiliania huxleyi* morphotypes in the Benguela coastal upwelling system (SE Atlantic). Mar. Ecol. Prog. Ser. *448*, 51-66.

Henriksen, K., Stipp, S.L.S., Young, J.R., and Marsh, M.E. (2004). Biological control on calcite crystallization: AFM Investigation of coccolith polysaccharide function, Amer. Mineral. *89*, 1709–1716.

Herfort, L., Thake, B., and Roberts, J. (2002). Acquisition and use of bicarbonate by *Emiliania huxleyi*. New Phytol. *156*, 427–436.

Hinz, D.J., Poulton, A.J., Nielsdóttir, M.C., Steigenberger, S., Korb, R.E., Achterberg, E.P., and Bibby, T.S. (2012). Comparative seasonal biogeography of mineralising nannoplankton in the Scotia Sea: *Emiliania huxleyi, Fragilariopsis* spp. and *Tetraparma pelagica*, Deep Sea Res. Part II, (59-60), 57–66.

Holligan, P.M., Groom, S., and Harbour, D.S. (1993). What controls the distribution of the coccolithophore, *Emiliania huxleyi*, in the North Sea? Fish. Oceanogr. *2*, 175–183.

Holtz, L.-M., Thoms, S., Langer, G., and Wolf-Gladrow, D.A. (2013). Substrate supply for calcite precipitation in *Emiliania huxleyi*: assessment of different model approaches. J. Phycol. *49*, 417-426.

Houdan, A., Billard, C., Marie, D., Not, F., Saez A.G., Young, J.R., and Probert, I. (2004). Holococcolithophore-heterococcolithophore (Haptophyta) life cycles: flow cytometric analysis of relative ploidy levels. Syst. Biodivers. *1*, 453-465

Houdan, A., Probert, I., Van Lenning, K., and Lefebvre, S. (2005). Comparison of photosynthetic responses in diploid and haploid life-cycle phases of *Emiliania huxleyi* (*Prymnesiophyceae*). Mar. Ecol. Prog. Ser. *292,* 139–146.

Huertas, I.E., Rouco, M., López-Rodas, V., and Costas, E. (2011). Warming will affect phytoplankton differently: evidence through a mechanistic approach. Proc. R. Soc. B. *278*, 3534-3543.

Hughs, J.S., and Otto, S.P. (1999). Ecology and the evolution of biphasic life cycles. Am. Nat. *154*, 306-320.

Iglesias-Rodriguez, M.D., Brown, C.W., Doney, S.C., Kleypas, J., Kolber, D., Kolber, Z., Hayes, P.K. and Falkowski, P.G. (2002). Representing key phytoplankton functional groups in ocean carbon cycle models: coccolithophorids. Glob. Biogeochem. Cy. *16*, 1100.

Iglesias-Rodriguez, M. D., Schofield, O.M., Batley, J., Medlin, L.K., and Hayes, P.K. (2006). Intraspecific genetic diversity in the marine coccolithophore *Emiliania huxleyi* (Prymnesiophyceae): the use of microsatellite analysis in marine phytoplankton population studies. J. Phycol. *42*, 526-536.

Iglesias-Rodriguez, M.D., Halloran, P.R., Rickaby, R.E.M., Hall, I.R., Colmenero-Hidalgo, E., Gittins, J.R., Green, D.R.H., Tyrrell, T., Gibbs, S.J, von Dassow, P., Rehm, E., Armbrust, E.V., and Boessenkool, K.P. (2008). Phytoplankton calcification in a high-CO_2 world. Science *320*, 336–340.

Iglesias-Rodriguez, M.D. (2013). Ocean acidification. In Earth System Monitoring, Orcutt, J, ed. (New York, U.S.A.: Springer) pp. 269-289.

Irie, T., Bessho, K., Findlay, H.S., and Calosi, P. (2010). Increasing costs due to ocean acidification drives phytoplankton to be more heavily calcified: optimal growth strategy of coccolithophores. PLOS ONE *5*, e13436.

Iwamoto, K., and Shiraiwa. Y. (2003). Characterization of NADH:Nitrate Reductase from the coccolithophorid *Emiliania huxleyi* (Lohman) Hay and Mohler (Haptophyceae). Mar. Biotechnol. *5*, 20 – 26.

Jones, B.M., Edwards, R.J., Skipp, P.J., O'Connor, C.D., and Iglesias-Rodriguez, M.D. (2011). Shotgun proteomic analysis of *Emiliania huxleyi*, a marine phytoplankton species of major biogeochemical importance. Mar. Biotechnol. *13*, 496–504.

Jones, B.M., Iglesias-Rodriguez, M.D., Skipp, P.J., Edwards, R.J., Greaves, M.J., Young, J.R., Elderfield, H., and O'Connor, C.D. (2013). Responses of the *Emiliania huxleyi* proteome to ocean acidification. PLOS ONE *8*, e61868

Jones, R.H., Flynn, K.J., and Anderson, T.R. (2002). Effect of food quality on carbon and nitrogen growth efficiency in the copepod *Acartia tonsa*. Mar. Ecol. Prog. Ser. *235*, 147-156.

Kaffes, A.,Thoms, S., Trimborn, S., Rost, B., Langer, G, Richter K.-U., Köhler, A., Norici, A., and Giordano, M. (2010). Carbon and nitrogen fluxes in the marine coccolithophore *Emiliania huxleyi* grown under different nitrate concentrations. J. Exp. Mar. Biol. Ecol. *393*, 1–8.

Kegel, J. U., John, U., Valentin, K., and Frickenhaus, S. (2013). Genome variations associated with viral susceptibility and calcification in *Emiliania huxleyi*. PLoS One *8*, e80684.

Klaveness, D., and Paasche, E. (1971). Two different *Coccolithus huxleyi* cell types incapable of coccolith formation. Arch. Mikrobiol. *75*, 382-385.

Klaveness, D. (1972). *Coccolithus huxleyi* (Lohm.) Kamptn. II. The flagellate cell. aberrant cell types, vegetative propagation and life cycles. Br. Phycol. J. *7*, 309–318.

Kolb, A., and Strom, S. (2013). An inducible antipredatory defense in haploid cells of the marine microalga *Emiliania huxleyi* (Prymnesiophyceae). Limnol. Oceanogr. *58*, 932–944.

Kroeker, K.J., Kordas, R.L., Crim, R.N., and Singh, G.G. (2010). Meta-analysis reveals negative yet variable effects of ocean acidification on marine organisms. Ecol. Letters *13*, 1419–1434.

Lakeman, M.B., von Dassow, P., and Cattolico, R.A. (2009). The strain concept in phytoplankton ecology. Harmful Algae *8*, 746-758.

Landry, D.M. Kristiansen, S., and Palenik, B.P. (2009). Molecular characterization and antibody detection of a nitrogen-regulated cell-surface protein of the coccolithophore *Emiliania huxleyi* (Prymnesiophyceae). J. Phycol. *45*, 650-659.

Langer, G., Nehrke, G., Probert, I., Ly, J., and Ziveri, P. (2009). Strain-specific responses of *Emiliania huxleyi* to changing seawater carbonate chemistry. Biogeosciences *6*, 2637-2646.

Langer, G., de Nooijer, L.J., and Oetjen, K. (2010). On the role of the cytoskeleton in coccolith morphogenesis: the effect of cytoskeleton inhibitors. J. Phycol *46*, 1252-1256.

Leblanc, K., Hare, C.E., Feng, Y., Berg, G.M., DiTullio, G.R., Neeley, A., Benner, I., Sprengel, C., Beck, A., Sanudo-Wilhelmy, S.A., Passow, U., Schreiber, K., Rowe, J. M., Wilhelm, S.W., Brown, C.W., and Hutchins, D.A. (2009). Distribution of calcifying and silicifying phytoplankton in relation to environmental and biogeochemical parameters during the late stages of the 2005 North East Atlantic Spring Bloom. Biogeosciences *6*, 2155–2179.

Lefebvre, S.C., Harris, G., Webster, R., Leonardos, N., Geider, R.J., Raines, C.A., Read, B.A., and Garrido, J.L. (2010). Characterization and expresson analysis of the Lhcf gene family in *Emiliania huxleyi* (Haptophyta) reveals differntial responses to light and CO_2. J. Phycol. *46*, 123-134.

Lefebvre, S.C., Benner, I., Stillman, J.H., Parker, A.E , Drake, M.K., Rossignol, P.E., Okimura, K.M., Komada, T., and Carpenter, E.J. (2012). Nitrogen source and pCO_2 synergistically affect carbon allocation, growth and morphology of the coccolithophore *Emiliania huxleyi*: potential implications of ocean acidification for the carbon cycle. Global Change Biol. *18*, 493-503.

Leonardos, N., and Geider, R.J. (2005). Elevated atmospheric carbon dioxide increases organic carbon fixation by *Emiliania huxleyi* (Haptophyta), under nutrient-limited, high-light conditions. J. Phycol. *41*, 1196-1203.

Leonardos, N., Read, B., Thake, B., and Young, J.R. (2009). No mechanistic dependence of photosynthesis on calcification in the coccolithophorid *Emiliania huxleyi* (Haptophyta). J. Phycol. *45*, 1046–1051.

Linschooten, C., van Bleijswijk, J.D.L., van Emburg, P .R., de Vrind, J.P.M., Kempers, E.S., Westbroek, P., and de Vrind-de Jong, E.W. (1991). Role of the light-dark cycle and medium composition on the production of coccoliths by *Emiliania huxleyi* (Haptophyceae). J. Phycol. *27*, 82-86.

Litchman, E., Klausmeier, C.A., Miller, J.R., Schofield, O.M., and Falkowski, P.G. (2006). Multi-nutrient, multi-group model of present and future oceanic phytoplankton communities. Biogeosciences *3*, 585-606.

Loebl, M., Cockshutt, A.M., Campbell, D.A., and Finkel, Z.V. (2010). Physiological basis for high resistance to photoinhibition under nitrogen depletion in *Emiliania huxleyi*. Limnol. Oceanogr. *55*, 2150–2160.

Lohbeck, K.T., Riebesell, U., and Reusch, T.B.H. (2012). Adaptive evolution of a key phytoplankton species to ocean acidification. Nat. Geosci. *5*, 346-351.

Mackinder, L., Wheeler, G., Schroeder, D., Riebesell, U., and Brownlee, C. (2010). Molecular mechanisms underlying calcification in coccolithophores. Geomicrob. J. *27*, 585–595.

Mackinder, L., Wheeler, G., Schroeder, D., von Dassow, P., Riebesell, U., and Brownlee, C. (2011). Expression of biomineralization-related ion transport genes in *Emiliania huxleyi*. Environ. Microbiol. *13*, 3250–3265.

Manton, I., and Leedale., G.F. (1969). Observations on the microanatomy of *Coccolithus pelagicus* and *Cricosphaera carterae*, with special reference to the origin and nature of coccoliths and scales. J. Mar. Biol. Ass. U. K. *49*, 1–16.

Marañon, E., Fernandez, E., Harris, R.P., and Harbour, D.S. (1996). Effects of the diatom-*Emiliania huxleyi* succession on photosynthesis, calcification and carbon metabolism by size-fractionated phytoplankton. Hydrobiologia *317*, 189-199.

Marsh, M.E. (1994). Polyanion-mediated mineralization - assembly and reorganization of acidic polysaccharides in the Golgi system of a coccolithophorid alga during mineral deposition. Protoplasma *177*, 108-122.

Marsh, M.E. (1999). Coccolith crystals of *Pleurochrysis carterae*: crystallographic faces, organization and development. Protoplasma *207*, 54-66.

Martinez, J.M., Schroeder D.C., and Wilson, W.H. (2012). Dynamics and genotypic composition of *Emiliania huxleyi* and their co-occurring viruses during a coccolithophore bloom in the North Sea. FEMS Microbiol. Ecol. *81*, 315–323.

Maynard Smith, J., and Price, G.R. (1973). The logic of animal conflict. Nature *246*, 15-18.

McKew, B.A., Davey, P., Finch, S.J., Hopkins, J., Lefebvre, S.C., Metodiev, M.V., Oxborough, K., Raines, C.A., Lawson, T., and Geider, R.J. (2013a). The trade-off between light-harvesting and photoprotective functions of fucoxanthin-chlorophyll proteins dominates light acclimation in *Emiliania huxleyi* (clone CCMP 1516). New Phytol. *200*, 74-85..

McKew, B.A., Lefebvre, S.C., Achterberg, E.P., Metodieva, G., Raines, C.A., Metodiev, M.V., and Geider, R.J. (2013b). Plasticity in the proteome of *Emiliania huxleyi* CCMP 1516 to extremes of light is highly targeted. New Phytol. *200*, 61-73.

Medlin, L.K., Barker, G.L.A., Campbell, L., Green, J.C., Hayes, P.K., Marie, D., Wriedens, S., and Vaulot, D. (1996). Genetic characterisation of *Emiliania huxleyi* (Haptophyta). J. Mar. Syst. *9*, 13−32.

Milliman, J.D. (1993). Production and accumulation of calcium carbonate in the ocean: budget of a nonsteady state. Glob. Biogeochem. Cy. *7*, 927–957.

Mohan, R., Mergulhao, L.P., Guptha, M.V.S., Rajakumar, A., Thamban, M., AnilKumar, N., Sudhakar, M., and Ravindra, R., (2008). Ecology of coccolithophores in the Indian sector of the Southern Ocean. Mar. Micropaleontol. *67*, 30–45.

Morse, J.W., Andersson, A.J., and Mackenzie, F.T. (2006). Initial responses of carbonate-rich shelf sediments to rising atmospheric pCO_2 and ocean acidification: role of high Mg-calcites. Geochim. Cosmochim. Acta *70*, 5814–5830.

Müller, M.N., Antia A.N., LaRoche, J. (2008). Influence of cell cycle phase on calcification in the coccolithophore *Emiliania huxleyi.* Limnol. Oceanogr. *53*, 506-512.

Müller, M.N., Schulz, K.G., and Riebesell, U. (2010). Effects of long-term high CO_2 exposure on two species of coccolithophores. Biogeosciences *7,* 1109-1116.

Müller, M.N., Beaufort, L., Bernard, O., Pedrotti, M.L., Talec, A., and Sciandra, A. (2012). Influence of CO_2 and nitrogen limitation on the coccolith volume of *Emiliania huxleyi* (Haptophyta). Biogeosciences *9*, 4155-4167.

Nanninga, H.J., and Tyrrell, T. (1996). Importance of light for the formation of algal blooms by *Emiliania huxleyi.* Mar. Ecol. Prog. Ser. *136* 195–220.

Nie, L., Wu, G., Culley, D.E., Scholten, J.C.M., and Zhang, W. (2007). Integrative analysis of transcriptomic and proteomic data: challenges, solutions and applications. Crit. Rev. Biotech. *27*, 63-75.

Nimer, N.A., Brownlee, C., and Merrett, M.J. (1994). Carbon dioxide availability, intracellular pH and growth irate of the coccolithophore *Emiliania huxleyi.* Mar Ecol. Prog. Ser. *109*, 257-262.

Nguyen, B., Bowers, R.M., Wahlund, T.M., and Read, B.R. (2005). Suppressive subtractive hybridization of and differences in gene expression content of calcifying and noncalcifying cultures of *Emiliania huxleyi* strain 1516. Appl. Environ. Microbiol. *71,* 2564-2575.

Otto, S.P., and Marks, J.C. (1996). Mating systems and the evolutionary transition between haploidy and diploidy. Biol. J. Linn. Soc. *57*, 197–218.

Paasche, E. (1962). Coccolith formation. Nature *193*, 1094-1095.

Paasche, E. (1964). A tracer study of the inorganic carbon uptake during coccolith formation and photosynthesis in the coccolithophorid *Coccolithus huxleyi*. Physiol. Plant. *3*, 1-82.

Paasche, E. (1965). The effect of 3(p-chlorophenyl)-1, 1-dimethylurea (CMU) on photosynthesis and light-dependent coccolith formation in *Coccolithus huxleyi*. Physiol. Plant. *18*, 138-145.

Paasche, E. (1966). Adjustment to light and dark rates of coccolith formation. Physiol. Plant. *19*, 271 -278.

Paasche, E. (2002). A review of the coccolithophorid *Emiliania huxleyi* (Prymnesiophyceae), with particular reference to growth, coccolith formation and calcification–photosynthesis interactions. Phycologia *40*, 503–529.

Palenik, B., and Henson, S.E. (1997). The use of amides and other organic nitrogen sources by the phytoplankton *Emiliania huxleyi*. Limnol. Oceanogr. *42*, 1544–1551.

Pedrotti, M.L., Fiorini, S., Kerros, M.-E., Middelburg, J.J., and Gattuso, J-P. (2012). Variable production of transparent exopolymeric particles by haploid and diploid life stages of coccolithophores grown under different CO_2 concentrations. J. Plankton Res. *34*, 1-11.

Poulton, A.J., Adey, T.R., Balch, W.M., and Holligan, P.M. (2007). Relating coccolithophore calcification rates to phytoplankton community dynamics: regional differences and implications for carbon export. Deep Sea Res. Part II. *54*, 538–557.

Poulton, A.J., Young, J.R., Bates, N.R., and Balch, W.M. (2011). Biometry of detached *Emiliania huxleyi* coccoliths along the Patagonian Shelf. Mar. Ecol. Prog. Ser. *443*, 1–17.

Poulton, A.J., Painter, S.C., Young, J.R., Bates, N.R., Bowler, B., Drapeau, D., Lyczsckowski, E., and Balch, W.M. (2013). The 2008 *Emiliania huxleyi* bloom along the Patagonian Shelf: Ecology, biogeochemistry, and cellular calcification. Glob. Biogeochem. Cy. *27*, 1023-1033.

Price, N.D., Papin, J.A., Schilling, C.H., and Palsson, B.O. (2003). Genome-scale microbial *in silico* models: the constraints-based approach. Trends Biotechnol. *21,* 162-169.

Qiu, Y.-L., Taylor, A.B., and McManus, H.A. (2012). Evolution of the life cycle in land plants. J. Syst. Evol. *50*, 171-194.

Quinn, P.S., Saez, A.G.,, Baumann, K-H., Steel, B.A., Sprengel, C., and Medlin, L.K. (2004). Coccolithophorid biodiversity: evidence from the cosmopolitan species *Calcidiscus leptoporus*. In Coccolithophores - From Molecular Processes to Global Impact, H.R. Thierstein and J.R. Young, eds. (New York, U.S.A.: Springer), pp. 299-326.

Quinn, P., Bowers, R.M., Zhang, X., Wahlund, T.M., Fanelli, M.A., Olszova, D. and Read, B.R. (2006). cDNA microarrays as a tool for identification of biomineralization proteins in the coccolithophorid *Emiliania huxleyi* (Haptophyta). Appl. Environ. Microbiol. *72*, 5512-5526.

Raffi, I., Backman, J., Fornaciari, E., Pälike, H., Rio, D., Lourens, L., and Hilgen, F. (2006). A review of calcareous nannofossil astrobiochronology encompassing the past 25 million years. Quaternary Sci. Rev. *25*, 3113-3137.

Ratti, S., Giordano, M., and Morse, D. (2007). CO_2-concentrating mechanisms of the potentially toxic dinoflagellate *Protoceratium reticulatum* (Dinophyceae, Gonyaulacales). J. Phycol. *43*, 693–701.

Read, B.A., Kegel, J., Klute, M.J., Kuo, A., Lefebvre, S.C., Maumus, F., Mayer, C., Miller, J., Monier, A., Salamov, A., Young, J., Aguilar, M., Claverie, J-M., Frickenhaus, S., Gonzalez, K., Herman, E.K., Lin, Y-C., Napier, J., Ogata, H., Sarno, A.F., Shmutz, J., Schroeder, D., de Vargas, C., Verret, F., von Dassow, P., Valentin, K., Van de Peer, Y., Wheeler, G., Allen, A.E., Bidle, K., Borodovsky, M., Bowler, C., Brownlee, C., Cock, J.M., Elias, M., Gladyshev, V.N., Groth, M., Guda, C., Hadaegh, A., Iglesias-Rodriguez, M.D., Jenkins, J., Jones, B.M., Lawson, T., Leese, F., Lindquist, E., Lobanov, A., Lomsadze, A., Malik, S-B., Marsh, M.E., Mackinder, L., Mock, T., Mueller-Roeber, B., Pagarete, A., Parker, M., Probert, I., Quesneville, H., Raines, C., Rensing, S.A., Riaño-Pachón, D.M., Richier, S., Rokitta, S., Shiraiwa, Y., Soanes, D.M., van der Giezen, M., Wahlund, T.M., Williams, B., Wilson, W., Wolfe, G.W., Wurch, L.L., Dacks, J.B., Delwiche, C.F., Dyhrman, S.T., Glöckner, G., John, U., Richards, T., Worden, A.Z., Zhang, X., and Grigoriev, I.V. (2013). Pan genome of the phytoplankton *Emiliania* underpins its global distribution. Nature *499*, 209-13.

Reid, E., Worthy, C.A., Probert, I., Ali, S.T., Love, J., Napier, J., Littlechild, J.A., Somerfield, P.J., and Allen, M.J. (2011). Coccolithophores: functional biodiversity, enzymes and bioprospecting. Mar. Drugs *9*, 586–60.

Reinfelder, J.R. (2011). Carbon concentrating mechanisms in eukaryotic marine phytoplankton. Ann. Rev. Mar. Sci. *3*, 291-315.

Richier, S., Kerros, M.-E., de Vargas, C., Haramaty, L., Falkowski, P.G., and Gattuso, J.-P. (2009). Light-dependent transcriptional regulation of genes of biogeochemical interest in the diploid and haploid life cycle stages of *Emiliania huxleyi*. Appl. Environ. Microbiol. *75*, 3366-3369.

Ridgwell, A., Schmidt, D.N., Turley, T., Brownlee, C., Maldonado, M.T., Tortell, P.D. and Young. J.R. (2009). From laboratory manipulations to Earth system models: scaling calcification impacts of ocean acidification. Biogeosciences *6,* 2611-2623.

Ridgwell, A., and Schmidt, D.N. (2010). Past constraints on the vulnerability of marine calcifiers to massive carbon dioxide release. Nat. Geosci. *3*, 196–200.

Riebesell, U., Zondervan, I., Rost, B., Tortell, P., Zeebe, R.E., and Morel, F. (2000). Reduced calcification of marine plankton in response to increased atmospheric CO_2. Nature *407*, 364-367.

Riebesell, U., Schulz, K.G., Bellerby, R.G.J., Botros, M., Fritsche, P., Meyerhöfer, M., Neill, C., Nondal, G., Oschliers, A., Wohler, J., and Zöllner, Z. (2007). Enhanced biological carbon consumption in a high CO_2 ocean. Nature *450*, 545–548.

Ries, J.B. (2010). Review: geological and experimental evidence for secular variation in seawater Mg/Ca (calcite-aragonite seas) and its effects on marine biological calcification. Biogeosciences *7*, 2795–2849.

Rokitta, S.D., de Nooijer, L.J., Trimborn, S., de Vargas, C., Rost, B., and John, U. (2011). Transcriptome analyses reveal differential gene expression patterns between the life-cycle stages of *Emiliania huxleyi* (Haptophyta) and reflect specialization to different ecological niches. J. Phycol. *47*, 829–838.

Rokitta, S.D., and Rost, B. (2012). Effects of CO_2 and their modulation by light in the life-cycle stages of the coccolithophore *Emiliania huxleyi*. Limnol. Oceanogr. *57*, 607–618.

Rost, B., Riebesell, U., Burkhardt, S., and Sültemeyer, D. (2003). Carbon acquisition of bloom-forming marine phytoplankton. Limnol. Oceanogr. *48*, 55-67.

Rouco, M., Branson, O., Lebrato, M., Iglesias-Rodríguez, M.D. (2013). The effect of nitrate and phosphate availability on *Emiliania huxleyi* physiology under different CO_2 scenarios. Front. Microbiol. *4*, 155.

Sciandra, A., Harlay, J., Lefèvre, D., Lemée, R., Rimmelin, P., Denis, M., Gattuso, J.-P. (2003). Response of coccolithophorid *Emiliania huxleyi* to elevated partial pressure of CO_2 under nitrogen limitation. Mar. Ecol. Prog. Ser. *26*. 111-122.

Schroeder, D.C., Biggi, G.F., Hall, M., Davy, J., Martinez, J.M., Richardson, A.J., Malin, G. and Wilson, W.H. (2005). A genetic marker to separate *Emiliania huxleyi* (Prymnesiophyceae) morphotypes. J. Phycol. *41*, 874-879.

Shiraiwa, Y. (2003). Physiological regulation of carbon fixation in the photosynthesis and calcification of coccolithophorids. Comp. Biochem. Physiol. B. Biochem. Mol. Biol. *136*, 775–783.

Siegel, H., Ohde, T., Gerth, M., Lavik, G., and Leipe, T. (2007). Identification of coccolithophore blooms in the SE Atlantic Ocean off Namibia by satellites and in-situ methods. Continental Shelf Research 27, 258–274.

Sikes, C.S., Roer, R.D., and Wilbur, K.M. (1980). Photosynthesis and coccolith formation: inorganic carbon sources and net reaction of deposition. Limnol. Oceanogr. *25*, 248−261.

Sikes, C.S., and Wheeler, A.P. (1982). Carbonic anhydrase and carbon fixation in coccolithophores. J. Phycol. *18*, 423-426.

Smith, H.E.K., Tyrrell, T., Charalampopoulou, A., Dumousseaud, C., Legge, O.J., Birchenough, S., Pettit, L.R., Garley, R., Hartman, S.E., Hartman, M.C., Sagoo, N., Daniels, C.J., Achterberg, E.P. and Hydes, D.J. (2012). Predominance of heavily calcified coccolithophores at low $CaCO_3$ saturation during winter in the Bay of Biscay. Proc. Natl. Acad. Sci. U.S.A., *109*. 8845–8849.

Sorrosa J.M., Satoh, M., and Shiraiwa, Y. (2005). Low temperature stimulates cell enlargement and intracellular calcification of coccolithophorids. Mar. Biotechnol. *7*, 128–133.

Suggett, D.J., le Floc'H, E., Harris, G.N. Leonardos., and Geider, R.J. (2007). Different strategies of photoacclimation by two strains of *Emiliania huxleyi* (Haptophyta). J. Phycol. *43*, 1209-1222.

Taylor, A.R., Russell, M.A., Harper, G.M., Collins, T.F.T., and Brownlee, C. (2007). Dynamics of formation and secretion of heterococcoliths by *Coccolithus pelagicus* sp. *braarudii*. Eur. J. Phycol. *42*, 125–136.

Taylor, J.R., and Stocker, R. (2012). Trade-offs of chemotactic foraging in turbulent water. Science *338*, 675-679.

Tettelin, H., Riley, D., Cattuto, C., and Medini, D. (2008). Comparative genomics: the bacterial pan-genome. Curr. Opin. Microbiol. *11*, 472-477.

Thierstein, H.R., Geitzenauer, K.R., Molfino, B., and Shackleton, N.J. (1977). Global synchroneity of late Quaternary coccolith datum levels: Validation by oxygen isotopes. Geology *5*, 400-404.

Trimborn, S., Langer, G., and Rost, B. (2007). Effect of varying calcium concentrations and light intensities on calcification and photosynthesis in *Emiliania huxleyi*. Limnol. Oceanogr. *52*, 2285-2293.

van Bleijswijk, J., van der Wal, P., Kempers, R., Veldhius, M., Young, J.R., Muyzer, G., de Vrind-de Jong, E. and Westbroek, P. (1991). Distribution of two types of *Emiliania huxleyi* (Prymnesiophyceae) in the northeast Atlantic region as determined by immunofluorescence and coccolith morphology. J. Phycol. *27*, 566-570.

van de Poll, W.H., Visser, R.J.W., and Buma, A.G. (2007). Acclimation to a dynamic irradiance regime changes excessive irradiance sensitivity of *Emiliania huxleyi* and *Thalassiosira weissflogii*. Limnol. Oceanogr. *52*, 1430–1438.

van der Wal, P., Jong, E.W., Westbroek, P., Bruijn, W.C., and Mulder-Stapel, A.A. (1983). Ultrastructural polysaccharide localization in calcifying and naked cells of the coccolithophorid *Emiliania huxleyi*. Protoplasma *118*, 157–168.

van der Wal, P., Johannes, P.M., de Vrind, E., de Vrind-de Jong, W., and Borman, A.H. (1987). Incompleteness of the coccosphere as a possible stimulus for coccolith formation in *Pleurochrysis carterae* (Prymnesiophyceae). J. Phycol. *23*, 218–221.

Van Mooy, B.A.S., Fredricks, H.F., Pedler, B.E., Dyhrman, S.T., Karl, D.M., Koblížek, M., Lomas, M.W., Mincer, T.J., Moore, L.R., Moutin, T., Rappé, M.S., and Webb, E.A. (2009). Phytoplankton in the ocean use non-phosphorus lipids in response to phosphorus scarcity. Nature *458*, 69-72.

Van Oostende, N, Moerdijk-Poortvliet, T.C.W., Boschker, H.T.S., Vyverman, W. and Sabbe, K. (2013). Release of dissolved carbohydrates by *Emiliania huxleyi* and formation of transparent exopolymer particles depend on algal life cycle and bacterial activity. Environ. Microbiol. *15*, 1514–1531.

von Dassow, P., Ogata, H., Probert, I., Wincker, P., Da Silva, C., Audic, S., Claverie, J.-M., and de Vargas, C. (2009). Transcriptome analysis of functional differentiation between haploid and diploid cells of *Emiliania huxleyi*, a globally significant photosynthetic calcifying cell. Genome Biol. *10*, R114.

von Dassow, P., and Montresor, M. (2011). Unveiling the mysteries of phytoplankton life cycles: patterns and opportunities behind complexity. J. Plankton Res. *33*, 3-12.

Wahlund, T.M., Hadaegh, A R., Clark, R., Nguyen, B., Fanelli, M., and Read, B. R. (2004). Analysis of expressed sequence tags from calcifying cells of marine coccolithophorid (*Emiliania huxleyi*). Mar. Biotechnol. *6*, 278-290.

Westbroek, P., Young, J., and Linschooten, K. (1989). Coccolith production (biomineralization) in the marine alga *Emiliania huxleyi*. J. Protozool. *36*, 368–373.

Westbroek, P., Brown, C.W., van Bleijswijk, J., Brownlee, C., Brummer, G.J., Conte, M., Egge, J., Fernandez, E., Jordan, R., Knappertsbusch, M., Stefels, J., Veldhuis, M., Van Der Wal, P., and Young, J.R. (1993). A model system approach to biological climate forcing. The example of *Emiliania huxleyi*. Global Planet. Change *8*, 27–46.

Wilbur, K.M., and Watabe, N. (1963). Experimental studies on calcification in molluscs and the alga *Coccolithus huxleyi*. Ann. N. Y. Acad. Sci. *109*, 82-112.

Winter, A., Jordan, R., and Roth, P.H. (1994). Biogeography of living coccolithophores in ocean waters. In Coccolithophores, A. Winter and W.G. Siesser, eds (Cambridge, U.K.: Cambridge University Press), pp. 161 – 177.

Winter, A., Henderiks, J., Beaufort, L., Rickaby, R.E.M., and Brown, C. W. (2014). Poleward expansion of the coccolithophore *Emiliania huxleyi*. J. Plankton Res. *36*: 316-325.

Xu, Y., Wahlund, T.M., Feng, L., Shaked, Y., and Morel F.M.M. (2006). A novel alkaline phosphatase in the coccolithophore *Emiliania huxleyi* (Prymnesiophyceae) and its regulation by phosphorus. J. Phycol. *42*, 835-844.

Young, J.R., Didymus, J.M., and Mann, S. (1991). On the reported presence of vaterite and aragonite in coccoliths of *Emiliania huxleyi*. Bot. Mar. *34*, 589-591.

Young, J.R., and Westbroek, P. (1991). Genotypic variation in the coccolithophorid species *Emiliania huxleyi*. Mar. Micropaleontol. *18*, 5-23.

Young, J.R., Davis, S.A., Bown, P.R., and Mann, S. (1999). Coccolith ultrastructure and biomineralization. J. Struct. Biol. *126*, 195–215.

Young, J.R., Geisen, M., Cros, L., Kleijne, A., Sprengel, C., Probert, I., and Østergaard, J.B. (2003). A guide to extant calcareous nannoplankton taxonomy. J. Nanno. Res. (Sp 1), 1−125.

Zeyl, C., Vanderford, T., and Carter, M. (2003). An evolutionary advantage of haploidy in large yeast populations. Science *299*, 555– 558.

Zondervan, I., Zeebe, R.E., Rost, B., and Riebesell, U. (2001). Decreasing marine biogenic calcification: a negative feedback on rising atmospheric *p*CO2. Glob. Biogeochem. Cy. *15*, 507-516.

Zondervan, I., Rost, B. and Riebesell, U. (2002). Effect of CO_2 concentration on the PIC/POC ratio in the coccolithophore *Emiliania huxleyi* grown under light-limiting conditions and different daylengths. J. Exp. Mar. Biol. Ecol. *272*, 55-70.

Zondervan, I. (2007). The effect of light, macronutrients, trace metals and CO_2 on the production of calcium carbonate and organic carbon in coccolithophores – A review. Deep Sea Res. Part II. *41*, 521–537.

Other Publications of Interest

Flores, E. and Herrero, A. (2014). The Cell Biology of Cyanobacteria. Caister Academic Press, U.K. ISBN: 978-1-908230-38-6

Herold, K.E. and Rasooly, A. (2009). Lab-on-a-Chip Technology: Fabrication and Microfluidics. Caister Academic Press, U.K. ISBN: 978-1-904455-46-2

Herold, K.E. and Rasooly, A. (2009) Lab-on-a-Chip Technology: Biomolecular Separation and Analysis. Caister Academic Press, U.K. ISBN: 978-1-904455-47-9

Wilkinson, M.G. (2015). Flow Cytometry in Microbiology: Technology and Applications. Caister Academic Press, U.K. ISBN: 978-1-910190-11-1

Xuefeng, L. (2014). Biofuels: From Microbes to Molecules. Caister Academic Press, U.K. ISBN: 978-1-908230-45-4

6

Bioluminescent Microalgae

Martha Valiadi[1]*, Charlotte L. J. Marcinko[2], Christos M. Loukas[2] and M. Débora Iglesias-Rodriguez[3]

[1]Electronics and Computer Science and Institute for Life Sciences, University of Southampton, Southampton, Hampshire, SO17 1BJ, U.K. M.Valiadi@soton.ac.uk
[2]National Oceanography Centre, Waterfront Campus, European Way, Southampton, SO14 3ZH, U.K. c.marcinko@noc.ac.uk
[3]Department of Ecology, Evolution and Marine Biology University of California Santa Barbara, Santa Barbara, CA 93106, U.S.A. iglesias@lifesci.ucsb.edu
*Author for correspondence.

Contents

Abstract

Some phytoplankton species, specifically members of the dinoflagellates, possess the remarkable ability to produce bioluminescence. Many bioluminescent species are globally distributed and some form blooms, which may sometimes be harmful. The bioluminescence system of dinoflagellates is unique from a biochemical, cellular and evolutionary perspective. Regulatory aspects of bioluminescence in relation to cell physiology remain largely unknown, as does the ecological niche of light-producing organisms. Meanwhile, in the field, bioluminescence has been used to study ecological dynamics within the plankton and to monitor toxic blooms. Studies on dinoflagellate bioluminescence have shed light on fundamental processes like cellular mechano-transduction, circadian rhythms and evolution of gene structure. The molecular components of the reaction have been used as reporters in biomedical applications and whole cells have been used to visualize fluid flow. Herein, we introduce key concepts and discuss current research that involves bioluminescent dinoflagellates. The research being conducted spans several different fundamental and applied fields, highlighting the vast benefits that dinoflagellate bioluminescence studies offer to biology and engineering.

General introduction

Dinoflagellates are the main eukaryotic protists that are capable of producing light (Widder, 2001; Haddock et al., 2010). Within this group, bioluminescence is present in a number of ecologically important and sometimes harmful species, many of which form blooms (Poupin et al., 1999; Valiadi et al., 2012). Indeed, dinoflagellates are responsible for most of the bioluminescence observed in the surface ocean (Tett, 1971), especially the globally known bright blue bioluminescent displays that have been reported since at least 500 BC (Harvey, 1957; Lynch, 1978).

The production of light occurs in dedicated organelles termed scintillons (DeSa and Hastings, 1968). Scintillons are dense vesicles approximately 0.5 – 0.9 µm in diameter (Johnson et al., 1985; Nicolas et al., 1987) and which, during the hours of darkness, are abundant in the periphery of the cell (Figure 6.1) (Fritz et al., 1990; Seo and Fritz, 2000). Scintillons contain the luciferin substrate, the luciferase enzyme (LCF) and, in most species, a luciferin binding protein (LBP) (Schmitter et al., 1976; Johnson et al., 1985; Knaust et al., 1998; Akimoto et al., 2004). Luciferin, the substrate of the reaction is also present in the scintillons. It is thought to originate by the breakdown of chlorophyll in photosynthetic species (Nakamura et al., 1989; Topalov and Kishi, 2001), although evidence suggests that there is more than one precursor molecule of luciferin in bioluminescent dinoflagellates (Valiadi and Iglesias-Rodriguez, 2013). Light is primarily produced in response to mechanical stimulation due to shear stress (Maldonado and Latz, 2007), for example upon contact with grazers or by breaking waves (Latz et al., 2008). However, many dinoflagellates also flash or glow spontaneously at certain times of the night (Krasnow et al., 1980; Colepicolo et al., 1993; Latz and Lee, 1995).

The primary bioluminescent taxonomic order is the Gonyaulacales, although the ability to produce light is present in a few other phylogenetically disparate dinoflagellate taxa (Poupin et al., 1999; Valiadi et al., 2012). Several bioluminescent species are cosmopolitan, in both coastal and open ocean regions, and include important heterotrophs (e.g. *Noctiluca* and *Protoperidinium*) and toxic (e.g. *Alexandrium*), or generally harmful species (e.g. *Noctiluca*, *Lingulodinium*, *and Ceratium*). There is considerable variation within genera and species in the presence of bioluminescence. For example, while almost all *Alexandrium* species are bioluminescent, two key oceanic genera, *Ceratium* and *Protoperidinium*, contain a very low proportion of bioluminescent representatives (Valiadi et al., 2012).

Function of bioluminescence in dinoflagellates

For bioluminescence to persist it must confer some competitive advantage. The ecological advantage of a bioluminescent lure to attract prey in organisms such as deep sea angler fish is obvious. The advantage provided by bioluminescence in single celled organisms such as dinoflagellates is less so. Wilson and Hastings (2013) hypothesized that bioluminescence systems initially evolved as oxygen detoxification systems in the

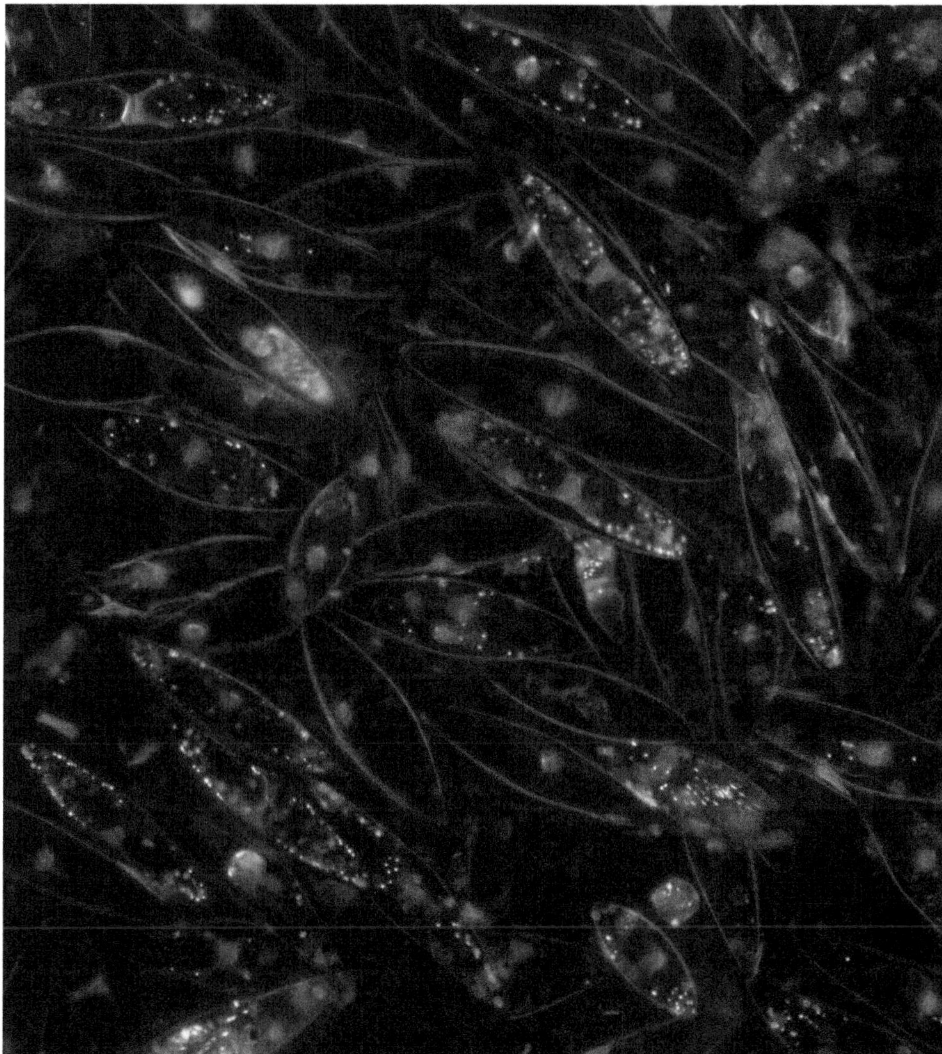

Figure 6.1. Microscopy image of cells of *Pyrocystis fusiformis* from a clonal stationary phase culture during the dark phase. Cells were visualized under an epifluorescence microscope at 10x magnification. The luciferin in the scintillons emits blue fluorescence when illuminated with ultraviolet light. In this picture individual scintillons appear as white speckles within the cells. Most are localized at the cell periphery and some are within cytoplasmic strands, which extend from the periphery, through the vacuole and towards the nucleus.

early Earth and then acquired an ecological function later on, when other oxygen detoxification mechanisms appeared. In dinoflagellates, it is now generally accepted that bioluminescence can provide an advantage by acting as an ecological survival strategy.

One theory put forward by which bioluminescence provides a survival benefit to dinoflagellates is the 'burglar alarm' hypothesis. Burkenroad (1943) proposed *"A peridinian sieved by a copepod might by its spark of light facilitate capture of the copepod by a herring... There might thus be a net advantage to the luminescent plankton from facilitation of the capture of primary predators..."*. Feeding activity by copepods on bioluminescent dinoflagellates is known to produce a bright spot of light lasting several seconds (Esaias and Curl, 1972) at a wavelength (~475 nm) which coincides with that required for maximum transmittance through water and the maximum visual sensitivity in many marine organisms (Tett and Kelly, 1973). Therefore, grazers feeding upon bioluminescent dinoflagellates are potentially highly visible to their predators.

The burglar alarm hypothesis has been supported by a number of laboratory studies which have indicated the ability of higher trophic organisms to utilise dinoflagellate bioluminescence to locate prey (Mesinger and Case, 1992; Fleisher and Case, 1995). Abrahams and Townsend (1993) found mortality rates of copepods increased by up to 70% in the presence of dinoflagellate bioluminescence and concluded that the observed increase in zooplankton mortality should translate into increased dinoflagellate survival. However, the study could not obtain cell density counts of dinoflagellates to support this conclusion and so could not unequivocally prove the survival benefit to the dinoflagellate population.

Bioluminescence may also provide an ecological advantage to dinoflagellates through a zooplankton 'startle response'. In this case, bioluminescent emission during close contact with a predator startles the predator and causes it to change course (Buskey et al., 1983) or suspend its feeding activity providing the dinoflagellate cell time to escape (Esaias and Curl, 1972). Experimental studies have shown that in the presence of dinoflagellate bioluminescence the ingestion rates of copepods can be reduced by 30 to 70% (Esaias and Curl, 1972; White, 1979). Bioluminescent flashes can cause changes in copepod swimming characteristics, increasing the number of high-speed swimming bursts and decreasing the amount of slow-speed swimming characteristic of their grazing behaviour (Buskey et al., 1983; Buskey and Swift, 1985). It has been inferred that the increased high-speed swimming activity represents a startle response by copepods which leads to a reduction in grazing. It is important to note that the hypothesized ecological advantages of dinoflagellate bioluminescence are not mutually exclusive. Indeed, there may be multiple ecological mechanisms by which bioluminescence can provide a survival benefit to dinoflagellates.

Cellular signaling cascade from stimulation to flashing

The chemical reaction to produce light is pH-dependent as LCF and LBP, which binds luciferin at physiological pH and become functional at ~ pH 6. At this pH structural changes in LCF make the luciferin binding sites available (Schultz et al., 2005) and LBP releases luciferin. Luciferin is oxidized by LCF to form oxyluciferin which results in the emission of photons (reviewed by Shimomura, 2006; Wilson and Hastings, 2013) in the form of a brief flash of blue light (with an approximate wavelength of 475 nm).

Translating a mechanical disturbance to a bioluminescence flash is achieved by a rapid a cascade of cellular processes (Figure 6.2). Shear stress exerted on the cell membrane

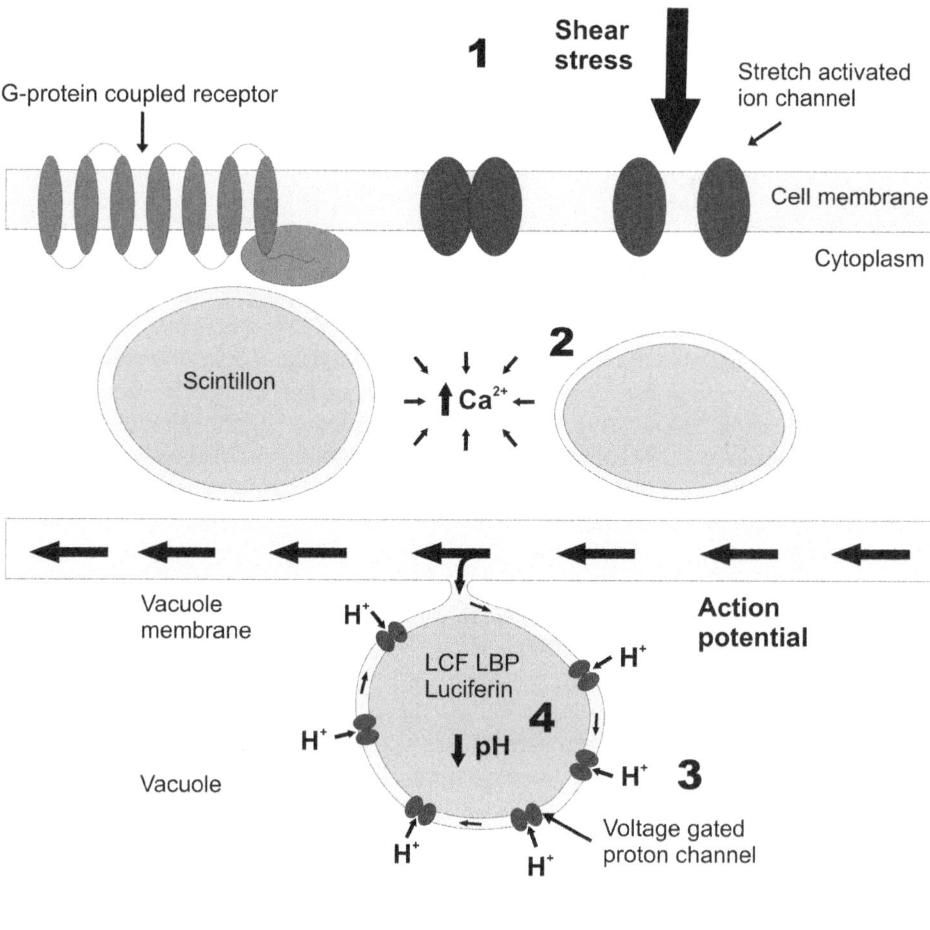

Figure 6.2. Schematic representation of part of a dinoflagellate cell depicting the cellular processes that take place to generate a bioluminescence flash. (1) Initially, a force is exerted on the outer cell membrane. Transduction of the signal from the outside to the inside of the cell involves the activation of a G-protein coupled receptor (Chen et al., 2007). The typical structure of such a receptor comprises seven transmembrane domains and the G-protein on the inner side of the cell membrane. The chemical signal is then transformed to an electrical signal on the vacuole membrane. (2) This is achieved by increasing the concentration of Ca^{2+} in the cytoplasm, mainly obtained from reserves within the cell and to a lesser extent from the external medium (von Dassow and Latz, 2002). This flux of Ca^{2+} causes the vacuole membrane to depolarize and generate an action potential. In this illustration, one of the scintillons protrudes into the vacuole while remaining attached to the vacuole membrane. This enables the action potential to travel to the scintillon membrane. (3) Voltage gated proton channels, which are presumably concentrated on the membrane of the scintillon rather than on the remaining vacuole membrane, are activated by the action potential allowing an influx of protons from the acidic vacuole. (4) This rapidly lowers the pH of the scintillons, activating the bioluminescent chemistry, and nearly instantly producing a flash of blue light. All the above processes, from the initial stimulus to the production of light, take place in just 20 milliseconds.

triggers a mechano-transduction pathway involving GTP-binding protein coupled receptors in the plasma membrane (Chen et al., 2007). Increase in cytosolic Ca^{2+}, mainly by release from intracellular stores (von Dassow and Latz, 2002) generates an action potential across the vacuole membrane and the associated scintillons. The action potential produced by these processes leads to an influx of protons from the acidic vacuole into the scintillons that protrude into the acidic vacuole, decreasing their internal pH from ~8 to ~6 (Fogel and Hastings, 1972). Voltage gated proton channels are involved in this process (Jin et al., 2013), presumably such as that recently characterized in another (non-bioluminescent) dinoflagellate (Smith et al., 2011). Recently, a role for stretch activated ion channels has also been revealed. These are activated by changes in the fluidity of the plasma membrane, for example by shear stress, which has now been shown to be the primary self-stimulator of spontaneous bioluminescence (Jin et al., 2013).

Molecular evolution

The evolution of dinoflagellate bioluminescence systems has been revealed through the sequencing and phylogenetic analyses of the dinoflagellate luciferase gene (*lcf*) and luciferin binding protein gene (*lbp*). The *lcf* in photosynthetic dinoflagellates is composed of three tandemly repeated domains (D1, D2 and D3) each consisting of a central region that is highly conserved at the amino acid level and encodes a catalytically active site (Li et al., 1997; Liu et al., 2004). This region is flanked by more variable N- and C-terminal regions with roles in the pH response and activity of the enzyme, respectively (Li et al., 2001; Suzuki-Ogoh et al., 2008). Four histidine residues in the N-terminal regions of each domain are thought to induce the pH-mediated conformational change that exposes the otherwise folded catalytic sites (Li et al., 2001; Schultz et al., 2005). An N-terminal gene region, which shows similarities to glutathione-S-transferase at the amino acid level, precedes the three catalytic domains in all studied species but has an unknown function (Li and Hastings, 1998; Liu et al., 2004).

Most photosynthetic dinoflagellates also contain *lbp*, which has now been fully or partially sequenced from five genera (Machabée et al., 1994; Hackett et al., 2005; Erdner and Anderson, 2006; Uribe et al., 2008; Jaeckisch et al., 2011; Valiadi and Iglesias-Rodriguez, 2014). The *lbp* of *L. polyedrum* occurs as a gene family with two distinct variants with approximately 86% identity (Lee et al., 1993; Machabée et al., 1994), each composed of four repeat domains with little similarity between them, unlike the homologous domains *lcf* (Liu and Hastings, 2007). Both variants are expressed at equal level (Machabée et al., 1994). We recently amplified *lbp* sequences with similar features from another four photosynthetic genera (Valiadi and Iglesias-Rodriguez, 2014).

An intriguing feature of the bioluminescence related genes of photosynthetic species is their evolution. Sequence comparisons of individual domains from triple domain *lcf* genes of different species have shown that corresponding domains of *lcf* are more similar in different organisms (e.g. D1 of *A. tamarense* and D1 of *P. reticulatum*) than are different domains within each organism (e.g. D1 and D2 of *A. tamarense*) (Liu et al., 2004). This is likely to reflect an ancient triplication of a *lcf* domain that was then carried forward during the evolution of the different photosynthetic species (Liu et al., 2004). However, the group of photosynthetic dinoflagellates that has been studied comprises very closely related species and it is therefore likely that much more variation exists, even within photosynthetic species. For instance, a highly divergent *lcf* gene exists within *Ceratium digitatum* (Valiadi et al., 2012) which is a representative of a cosmopolitan open ocean bioluminescent dinoflagellate genus (Swift et al., 1995). Further divergence among sequences arises from the multiplications of gene copies in dinoflagellate genomes. Both *lcf* and *lbp* phylogenies have shown that the divergence of sequences between different gene copies may exceed the divergence of sequences from different organisms (Valiadi et al., 2012; Valiadi and Iglesias-Rodriguez, 2014). Therefore, both genes appear to be under strong selection pressure leading to rapid continuous mutations.

The complexity of the evolution of dinoflagellate bioluminescence system is highlighted by the unique bioluminescence gene of *N. scintillans* (Liu and Hastings, 2007), a heterotrophic dinoflagellate which is ancestral to photosynthetic bioluminescent species. Its *lcf* consists of only one domain which is shorter than that of photosynthetic species and it is attached to *lbp* as a single hybrid gene (Figure 6.3). The N-terminal gene region lacks any sequence similarity and is much shorter than that of photosynthetic species. Additionally, the *lcf* domain is shorter by 60 AA at the N-terminal than the *lcf* domains of

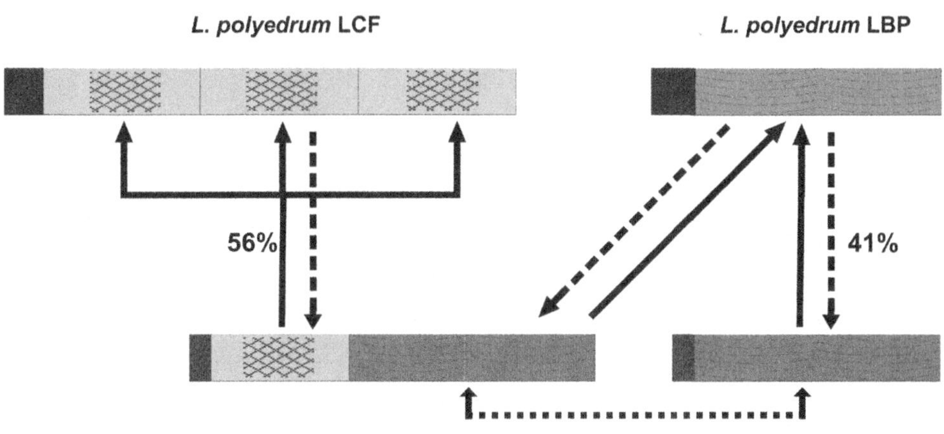

Figure 6.3. A representation of the different LCF and LBP structures, per cent identities between gene domains and the proposed scenarios for their evolution. *Lingulodinium polyedrum* (Lp) LCF represents all LCF's fully characterized from photosynthetic dinoflagellates to date (Liu et al., 2004). Lp LCF contains the three non-identical repeated domains, each 377 AA long with 150 AA in the central regions that encode the catalytic sites (patterned) and retain higher sequence similarity. The three domains are preceded by an N-terminal gene region. *L. polyedrum* also contains an LBP, which consist of 4 repeated domains and a similar N-terminal gene region to Lp LCF. In *Noctiluca scintillans* (Ns), LCF and LBP exist as a single gene. It contains only one LCF domain which is most similar to the second domain of Lp LCF. The LBP domain in Ns is structurally equivalent to that in Lp. The N-terminal gene region preceding the LCF and LBP in Ns is not structurally equivalent to that which precedes the analog genes in Lp. Two potential scenarios of how these gene structures arose are shown. In the gene fusion scenario (dashed lines), the second domain of LCF was excised (e.g. by splicing) and fused with LBP in *N. scintillans*. It is currently impossible to tell whether the solitary LBP in *N. scintillans* preceded the domain fused to LCF, or vice versa (dotted line). In the gene fission scenario (solid line), the Ns gene split and the LCF domain underwent successive duplications. Again, whether the Lp LBP could be derived from the fused LBP domain or the solitary LBP gene, is unknown.

photosynthetic species and is thus missing 3 of the 4 histidines that are thought to be responsible for pH regulation (Li et al., 2001), although the pH regulation remains. Despite these major structural differences, both the functionally relevant central part of the *lcf* domain and *lbp* domain retain sequence similarity at the amino acid level to *L. polyedrum*. A likely scenario is that the *N. scintillans lcf/lbp* could have undergone fission giving rise to the separate genes in photosynthetic species. The single *lcf* domain could have then triplicated in photosynthetic species to produce a protein with a triple catalytic capacity and hence a higher bioluminescence yield. Further evidence for such an evolutionary event is supported by a solitary *lbp* domain in *N. scintillans* which is not attached to *lcf*, but rather directly to the N-terminal gene region (Valiadi and Iglesias-Rodriguez, 2014). Now it is important to determine whether there is also a solitary *lcf* in *N. scintillans*, which could indicate fission in some of the *lcf/lbp* gene copies, or if the solitary *lbp* represents a splice variant of the original hybrid gene. Answering these questions is important to reveal the selective pressures that drive the evolution of bioluminescence through time. For example, gene fission in *N. scintillans* could arise from the need to differentially regulate the expression of *lcf* and *lbp*, allowing the enzyme to be recycled while the binding protein can remain proportional to the substrate pool. A triplication of *lcf* domains and therefore a triplication of catalytic capacity could signify the need to maintain bioluminescence intensity while reducing the number of scintillons that can be contained in these smaller cells (Valiadi and Iglesias-Rodriguez, 2013).

Physiology

Diet effects on bioluminescence

The relationship between bioluminescence and cell metabolism and overall physiology has not been investigated in detail. Nutrient availability appears to be linked to bioluminescence although the exact mechanism of action is unknown. In general, when cultures of photosynthetic species become nutrient-limited (Esaias et al. 1973, Sweeney 1981) bioluminescence is diminished. However, in the photosynthetic species *Alexandrium fundyense*, there are indications that the expression of *lbp* is higher under phosphate limitation than under nitrate limitation (Erdner and Anderson 2006). This suggests that more luciferin is produced under these conditions but it is unknown whether the effect on bioluminescence is proportional. In heterotrophic organisms, bioluminescence may also diminish when cells are starved (Buskey et al. 1994, Buskey et al. 1992, Li et al. 1996). Several stressors can alter dinoflagellate physiology in a distinctive way. For example, in *Protoperidinium depressum*, the bioluminescence system became even more sensitive to photoinhibition after a period of starvation (Li et al. 1996). In the species *P. divergens* and *P. crassipes*, bioluminescence was correlated with feeding frequency rather than with growth rate and cells continued to invest energy in bioluminescence despite being unable to grow (Latz and Jeong, 1996). These findings highlight the need to decipher the actual mechanisms linking bioluminescence and other physiological processes and the importance of studying the effect of multiple stressors on dinoflagellate bioluminescence.

Physiological effects of ocean acidification

Recent concern about one potential stressor to phytoplankton, ocean acidification (Caldeira and Wickett, 2003), has prompted efforts focused primarily on calcium carbonate-producing species (coccolithophores) (see Iglesias-Rodriguez, 2013). We report on the effect of ocean acidification on dinoflagellate physiology and specifically, on the effects on bioluminescence of the photosynthetic species *Lingulodinium polyedrum*. In these experiments, cells of exponentially-growing *L. polyedrum* were exposed to CO_2 partial pressures representative of near-present day (385 µatm CO_2) and future (750, 1 200 and 1 500 µatm CO_2) climate scenarios for ten days, over 2-3 generations. This is within the time frame of expected responses to ocean acidification in phytoplankton; for example, Barcelos e Ramos et al. (2010) showed that the coccolithophore species *Emiliania huxleyi* acclimated in the short-term (26 h) after cultures grown at 500 µatm CO_2 were abruptly exposed to 190, 410, 800 and 1 500 µatm CO_2. Cultures were bubbled with air containing CO_2 partial pressures of 350, 750, 1 200 and 1 500 µatm and a prior test using 385 µatm CO_2 confirmed that there was no significant difference in cell growth and bioluminescence between the bubbled and the non-bubbled cultures (not shown).

Growth rates in cells exposed to 1 500 µatm CO_2, appeared to require a more pronounced lag phase before reaching growth rates comparable to those of cultures exposed to lower CO_2 conditions (Figure 6.4). These results are consistent with studies suggesting a correlation between dinoflagellate abundance and elevated seawater pH/ low CO_2 conditions (Yoo, 1991; Hinga, 1992). Although photosynthetic rates are not available from this study, the increasing trend in O_2 concentration in the media over 10 days suggests a general increase in photosynthetic carbon fixation with increasing CO_2 (Table 6.1). Specifically, results showed a slight (<10%) decline in O_2 concentration from the 350 to the 750 µatm CO_2 treatment and an increase in $\Delta O_2/\Delta t$ at 1 200 and 1 500 µatm CO_2, suggesting that increasing CO_2 is favorable to photosynthetic carbon fixation.

Observations of cellular bioluminescence indicated significant differences in average bioluminescence values between the different CO_2 treatments and similar non-uniform responses to the different conditions with time (Figure 6.4). Specifically, results revealed an increase followed by a decline in bioluminescence, with maximum values being reached on days 2 (385 µatm CO_2) and 6 (750, 1 200 and 1 500 µatm CO_2) of growth (Figure 6.4). The lowest values towards the end of the experiment were observed in the lowest CO_2 treatment (Figure 6.4) but a link between the magnitude of bioluminescence and the CO_2 levels used in the treatments could not be identified.

Table 6.1. Carbonate chemistry and O_2 data of *L. polyedrum* cultures. For each parameter the first value represents day 0 and the second number represents day 10 of the experiment.

Treatment [CO_2] (µatm)	pH	pCO_2 (µatm)	HCO_3^- (µmol/kg)	CO_3^{2-} (µmol/Kg)	CO_2 (µmol/Kg)	O_2 (µmol/kg)
385	8.66/ 10.05	143.6/ 1.1	1772.4/ 298.9	254.3/ 1029.0	5.3/ 0.0	217.6/ 267.0
750	8.38/10. 01	300.5/ 1.2	1958.2/ 328.4	149.6/ 1048.9	11.1/ 0.0	230.7/ 276.9
1200	8.27/ 9.88	397.2/ 2.3	2017.0/ 439.3	119.1/ 1021.1	14.6/ 0.1	224.2/ 280.2
1500	8.26/ 9.16	417.7/ 34.3	2072.0/ 1300.5	119.5/ 589.2	15.4/ 1.3	222.5/ 303.3

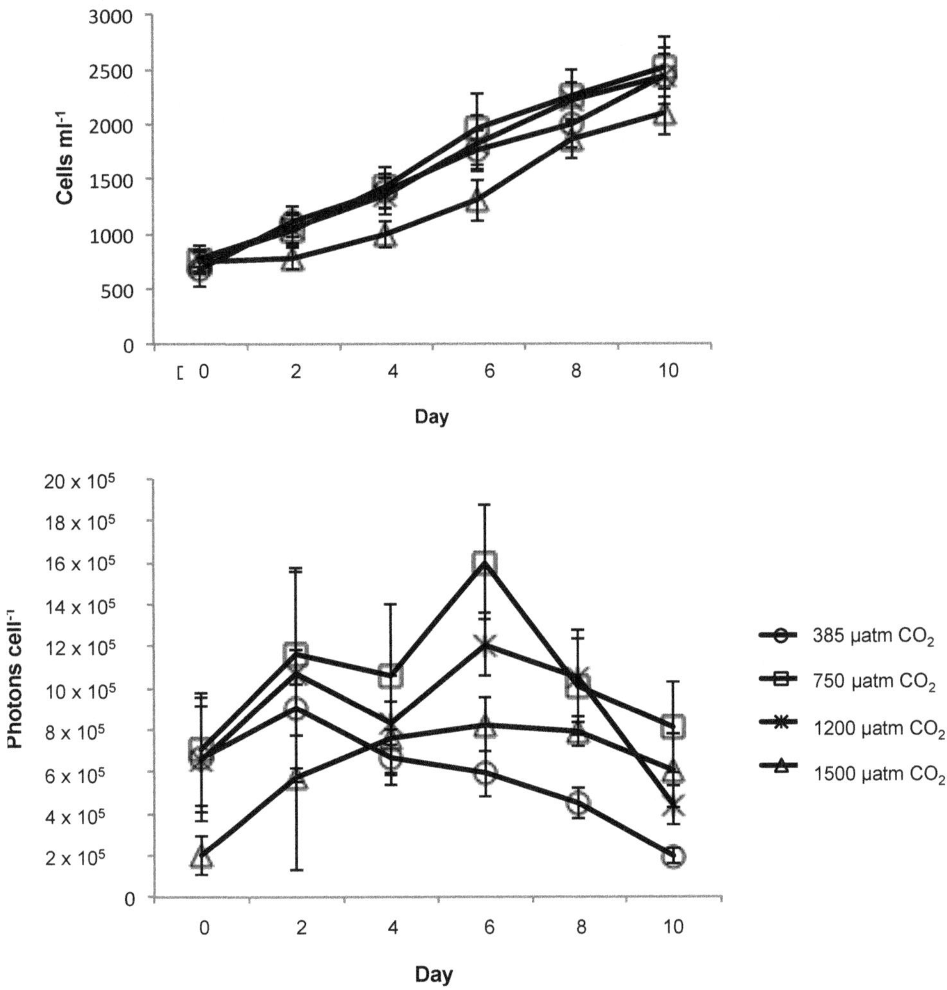

Figure 6.4. Plots of growth curved of *L. polyedrum* cells exposed to different concentrations of CO_2. Growth is expressed in cells per unit volume in (A) and correlated to cellular bioluminescence in (B).

In the two lowest CO_2 treatments, cells are likely to have been CO_2 limited rather than nitrate or phosphate limited as, by the end of the experiment, both NO_3^- and PO_4^{2-} were still available in the media in all treatments (results not shown) but CO_2 was undetectable in cultures maintained at 385 and 750 µatm CO_2. Indeed, dinoflagellates appear to possess a moderately efficient carbon concentrating mechanism, which is particularly important given their very low CO_2 affinity of their form II ribulose-1,5-bisphosphate carboxylase/oxygenase (RuBisCO) (Dason et al. 2004; Lapointe et al. 2008; Reinfelder, 2011). However, there is an example revealing that the non-bioluminescent species *Protoceratium reticulatum* (Poupin *et al.* 1999) lacks carbonic anhydrase external to the plasmamembrane and that this species does not utilize bicarbonate ions (HCO_3^-) (Ratti et al. 2007). Therefore, an increase in CO_2 could represent an energetic advantage in maintaining photosynthetic carbon fixation in species without an efficient carbon concentrating mechanism.

It is well established that dinoflagellate growth rates can be enhanced when seawater pH exceeds 8.5 (Hinga, 1992), although it is known that there are significant physiological differences between phylotypes/strains of the same species in dinoflagellates (e.g., Brading et al., 2011). This phenotypic as well as genotypic diversity of inherent properties and responses to environmental stimuli in dinoflagellates (Frommlet and Iglesias-Rodriguez, 2008; Alpermann et al., 2010) is actually observed across phytoplankton taxa; for example, in coccolithophores (Iglesias-Rodriguez et al., 2006; Langer et al., 2009) and in diatoms (e.g., Rynearson et al., 2006; Vanormelingen et al., 2008). This diversity suggests a potential for population resilience in bioluminescent dinoflagellate populations to environmental pressure. The synergistic effects of environmental variables also suggests that ocean acidification and other environmental factors (e.g., alteration in nutrient loading) will alter the relative abundance of species and therefore the associated biogeochemical fluxes in the future.

Photoenhancement of bioluminescence

Laboratory studies using cultures of *Lingulodinium polyedrum* and *Ceratium fusus* have found significant positive relationships between organisms' bioluminescence capacity and previous daytime irradiance (Sweeney, 1981; Sullivan and Swift, 1995). Such responses signify photo-enhancement of dinoflagellate bioluminescence and indicate that the short term light history of organisms may be an important influence on night-time bioluminescence intensity. However, evidence for photo-enhancement in natural dinoflagellate populations is mixed and not well defined due to lack of observations.

Within the subpolar North Atlantic (59° N, 21° W) during summer 1991, the bioluminescence capacity of *Ceratium fusus* showed no evidence of photo-enhancement. Comparisons indicated that cells from the dimly lit thermocline demonstrated a greater bioluminescence capacity than those within the well-lit mixed layer (Sullivan and Swift, 1995). Conversely, experiments conducted on natural mixed populations further south at the Porcupine Abyssal Plain (49° N 16° W) during summer 2009 indicated that exposure to daytime irradiance was required to maintain night-time bioluminescence and that maximum night-time bioluminescence intensity increased monotonically with previous daytime irradiance (Marcinko et al., 2013). These results suggest that the daytime irradiance received by the mixed bioluminescent populations had a direct influence upon the stimulated night-time bioluminescence intensity observed and support laboratory findings of photo-enhancement.

It is currently unclear whether photo-enhancement is restricted to specific bioluminescent species and the mechanism by which bioluminescence can be photo-enhanced is unknown. Further laboratory and field studies are required to fully understand the effects of irradiance on dinoflagellate bioluminescence systems, how common this is in natural populations and how it may vary spatially and seasonally.

Diel rhythm of bioluminescence

Studies of diel rhythm regulation in laboratory cultures

Most bioluminescent dinoflagellates display a diel rhythm in bioluminescence intensity, being much brighter in the night than in the day. The two main mechanisms underlying this variation are the regulation of bioluminescence at the molecular and cellular level by a circadian clock and/or photoinhibition of bioluminescence in the day.

In photosynthetic dinoflagellates, bioluminescence exhibits a diel rhythm controlled by an endogenous circadian clock. This makes bioluminescence almost undetectable in the day and brightest at night when it is most effective. The regulation of bioluminescence circadian rhythms varies significantly in the two dinoflagellate species (*L. polyedrum* and *P. lunula*) that have been investigated in depth (reviewed extensively by Hastings, 2007). In *L. polyedrum* the quantity of LCF and LBP is regulated translationally, while mRNA levels of both genes remain stable throughout the light-dark cycle (Morse et al., 1989; Knaust et al., 1998; Mittag et al., 1998). It has been suggested that this process is mediated by a repressor protein which binds to the 3' untranslated region of the *lcf* mRNA (Mittag et al., 1994), however, this has been disputed (Lapointe and Morse, 2008). The LCF, LBP, luciferin and even the whole scintillons are destroyed at dawn and then begin to be resynthesized at dusk; they peak in quantity approximately four hours into the night when they reach amounts that are 10 times higher than in the day (Dunlap and Hastings, 1981; Johnson et al., 1984; Fritz et al., 1990; Akimoto et al., 2004). In contrast, *P. lunula* regulates its bioluminescence in a completely different way. The number of scintillons and their luciferin and LCF content, as well as *lcf* mRNA, do not vary (Knaust et al., 1998; Okamoto et al., 2001). Instead, the scintillons are relocated concurrently with the chloroplasts to modulate bioluminescence intensity, placing the scintillons at the periphery of the cell during the night but near the centre of the cell during the day to prevent their stimulation (Seo and Fritz, 2000). These positional changes involved both actin filaments and microtubules of the cytoskeleton (Heimann et al., 2009). Recently, Hastings (2013) outlined a hypothesis for role of synthesis and destruction of proteins. He presented evidence indicating the re-allocation of amino acids in different proteins at different times of day in *L. polyedrum*. In this way, nitrogen can be conserved in an environment where it can be limiting (e.g. in a bloom). The lack of destruction of bioluminescence proteins in *Pyrocystis* can be justified by a rich nutrient supply in deeper water normally inhabited by this organism (Swift and Meunier, 1976). These findings are consistent with observations made by Roy *et al.* (2014) where cells of *L. polyedrum* that formed cold-induced cysts lowered their metabolic rate and arrested their bioluminescence circadian rhythm.

Bioluminescence intensity in most dinoflagellates is diminished in the day by photoinhibition (Sweeney et al., 1959; Esaias et al., 1973; Hamman et al., 1981b; Hamman and Seliger, 1982; Sullivan and Swift, 1994; Li et al., 1996). This is unrelated to whether or not a dinoflagellate also controls its bioluminescence by a circadian rhythm as it does not depend on the concentration of available substrate of protein for bioluminescence. Rather, the response to the mechanical stimulation is blocked, presumably by a change in the polarization of the cell membrane (Hamman and Seliger, 1982). In most bioluminescent species, maximal photoinhibition is caused by blue light (Esaias et al., 1973; Hamman et al., 1981a; Sullivan and Swift, 1994; Li et al., 1996). The identity of the receptor of the inhibiting light is unknown, as are the cellular processes that bring about photoinhibition of bioluminescence. Nevertheless, limiting the energy expenditure associated with bioluminescence only to the times when it is visible is an important evolutionary adaptation.

Diel bioluminescence rhythm in natural dinoflagellate populations
Diel variations in bioluminescence potential have often been observed in field studies (Figure 6.5). Increases of several orders of magnitude between day and night have been recorded in many regions including the Southern California Bight, Black Sea and Vestfjord (Norway) (Greenblatt et al., 1984; Lapota et al., 1992a; Tokarev et al., 1999b). Measurements compiled across several regions of the world's oceans show night-time bioluminescence intensity in surface waters is typically one to three orders of magnitude greater than corresponding daytime intensities (Utyushev et al., 1999). Early explanations for this variability were attributed to changes in the composition of bioluminescent organisms being measured (Greenblatt et al., 1984; Lapota and Losee, 1984) or the vertical migration of organisms through the water column (Seliger et al., 1962). However, variations are now known to be more often associated with endogenous circadian regulation or the exogenous diel rhythm of dinoflagellate bioluminescence.

Within the Sargasso and Mediterranean Seas photoinhibition was thought to be responsible for significant increases in the number of flashes and intensity measured from day to night, which could not be explained by changes in the vertical distribution of organisms (Batchelder et al., 1992; Cussatlegras et al., 2001). In Vestfjord, Norway, Lapota et al. (1992b) demonstrated that the diel changes in measured bioluminescence were caused by photoinhibition of naturally occurring *Ceratium fusus* and *Protoperidinium curtipes* which were found to increase their night-time flash intensities by 137 and 29 times respectively relative to daytime values.

A small number of field studies have detected circadian regulation of bioluminescence within natural dinoflagellate communities. These include within populations of *Pyrodinium bahamense* within Oyster Bay, Jamaica (Soli, 1966; Biggley et al., 1969) and mixed communities of bioluminescent species from coastal salt ponds in Woods Hole (Kelly and Katona, 1966). In Woods Hole, detection of circadian regulation was not persistent year round and was more generally found to be associated with the seasonal presence of photosynthetic *Gonyaulax* spp. (Kelly, 1968; Yentsch and Laird, 1968). Similarly, evidence of circadian regulation in mixed dinoflagellate communities within the North Atlantic has also been attributed to the presence of photosynthetic *Gonyaulax* and *Pyrocystis* species (Figure 6.6, Marcinko et al., 2013). Differences in diel rhythms between species can make variations in bioluminescence within natural multi-species plankton communities complex and dependent on community structure (Utyushev et al., 1999).

Applications

Technological applications
Dinoflagellate bioluminescence and luciferase has recently been used in several applied sciences. The earliest technological use of bioluminescent dinoflagellates was the QwikLite toxicity assay which used several dinoflagellates as a reporter for toxic metals

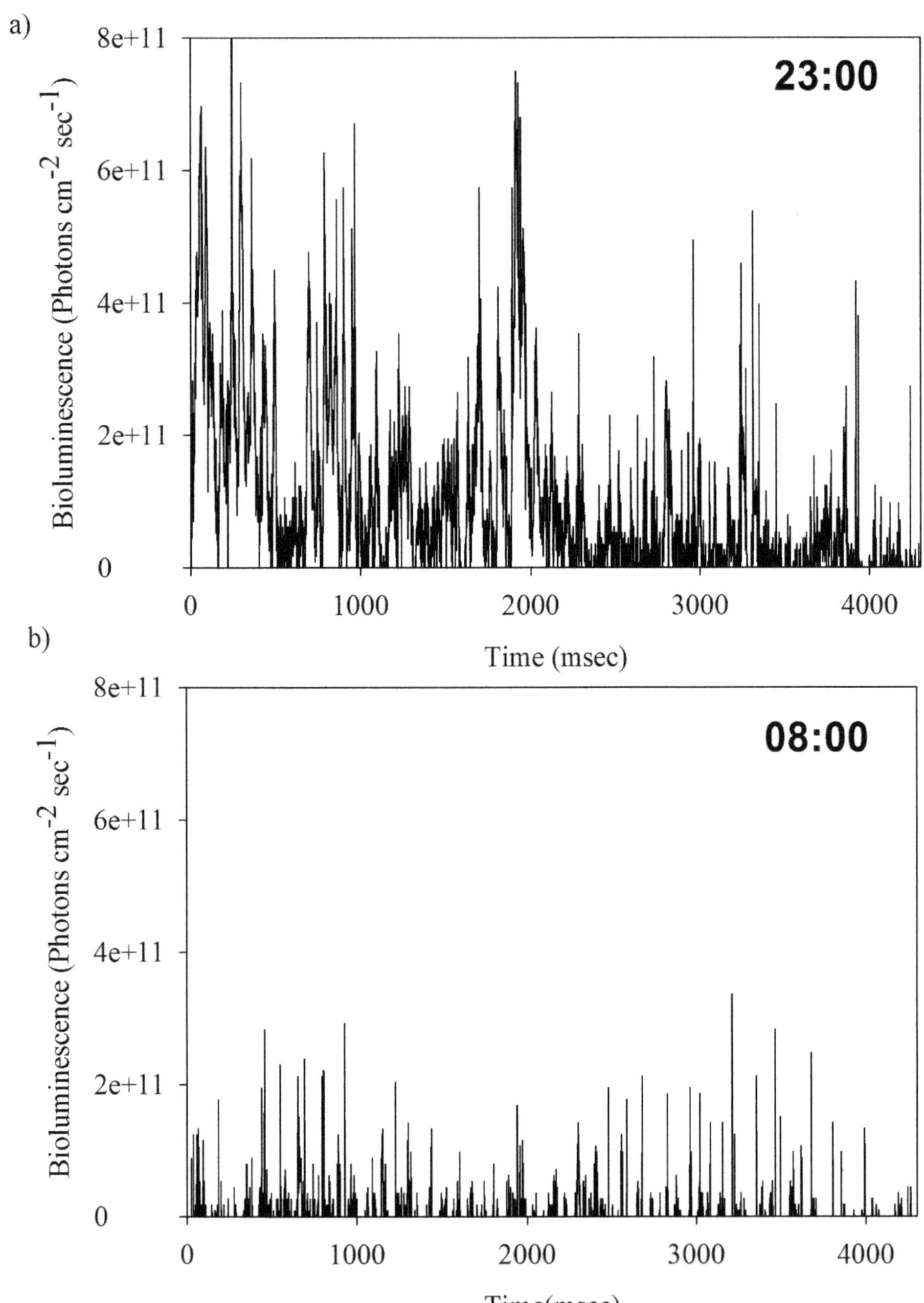

Figure 6.5. Example of bioluminescence potential stimulated from a mixed community of bioluminescent dinoflagellates collected from 5 m depth at the Porcupine Abyssal Plain observatory in the North Atlantic. Samples of 2L volume were measured using a GLOWtracka bathyphotometer adapted for bench-top use (see Marcinko et al. 2013 for details). Bioluminescence was stimulated via flow through a 1 mm mesh excitation grid. Measurements were taken at a) night (23:00 hours) and b) during the day (08:00 hours).

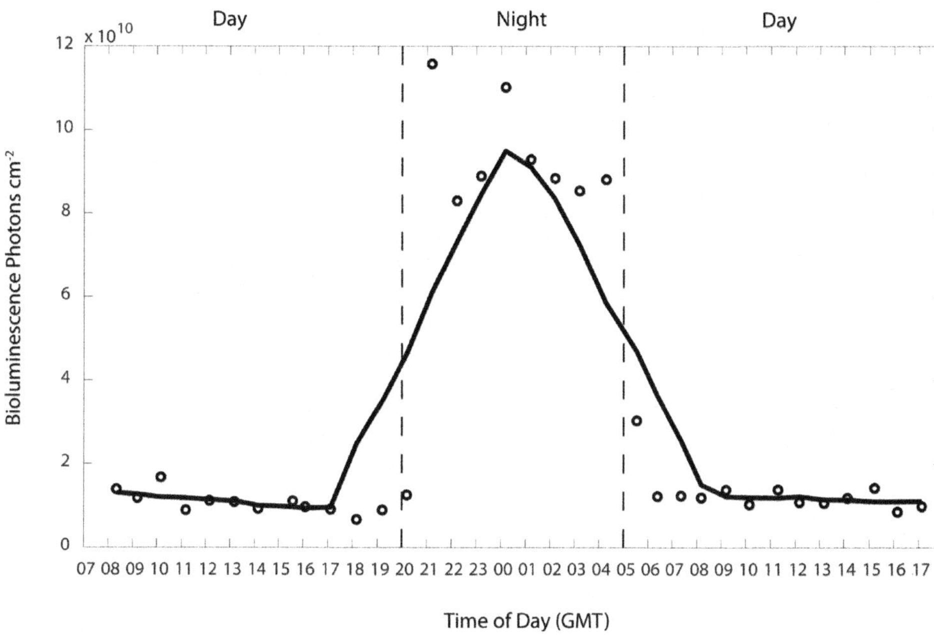

Figure 6.6. Example of diel variability in bioluminescence potential stimulated from a community of bioluminescent dinoflagellates dominated by photosynthetic species including Gonyaulax and Pyrocystis species. Seawater (120 L) was collected from 5-meter depth at the Porcupine Abyssal Plain observatory in the North Atlantic during August 2009. Samples were incubated *in situ*, at a 16°C, and exposed to the natural photoperiod. A volume of 2 L was measured hourly using a GLOWtracka bathyphotometer adapted for bench-top use and bioluminescence was stimulated via flow through a 1-mm mesh excitation grid. (Adapted from Marcinko et al 2013).

and organic pollutants in seawater (Lapota et al., 2007). Here, the diminishing of cellular bioluminescence was used to derive a toxicity index (LD50) to various compounds and samples of contaminated sediments, and compared with other ecotoxicological assays. Dinoflagellate bioluminescence was found to be a highly sensitive and cost-effective approach to detect of toxicity in the marine environment. Dinoflagellate luciferase has now also been used as a reporter in biomedical research, especially in cell biology. Single domains of dinoflagellate LCF fused to genes of interest in mammalian cells produce a fusion protein whose abundance can be monitored by a rapid bioluminescence assay, thus allowing for monitoring gene expression levels in the tagged genes (Suzuki et al., 2005). In a subsequent study, Kato et al. (2011) used dinoflagellate LCF to monitor the expression and kinetics of a cell membrane protein. The authors fused LCF to the extracellular part of platelet-derived growth factor receptor transmembrane protein of a mammalian cell, which could then be detected by the addition of luciferin. The extracellular part could be specifically monitored by adding luciferin, which does not penetrate the cell membrane. The intracellular part could be also targeted separately by first quenching the extracellular part using sulfo-NHS-acetate and then lysing the cells to expose the intracellular fraction to the added luciferin. Separate identification of cellular fractions allow for targeted studies on membrane trafficking and for testing the effects of pharmaceutical drugs. For example, Li et al (2010) used dinoflagellate luciferase to investigate treatments of beta-galactosidosis, a condition leading to severe neuronal dysfunction in humans. Luciferase was used to monitor the degree of restoration of the enzyme activity of mutated beta-galactosidase after treatment with a chemical chaperone which stabilized the enzyme. High-throughput screening by means of a bioluminescence assay allows for effective testing of novel compounds for treatments in disease.

Another key use of dinoflagellate bioluminescence is in fluid mechanics. This has particular relevance for coastal engineering where accurate knowledge of a flow field guides the design of constructions like wave-breakers and erosion protection structures on beaches. The force field within fluids can typically only be visualized by contact on measuring devices, which themselves cause changes to the flow field. The incredible sensitivity of cells to shear stress and the magnitude of the light produced make them particularly useful as reporters of fluid forces without contact instrumentation. Two recent studies have explored the use of bioluminescent dinoflagellates to visualize and quantify the flow field associated with the flow of water of a rippled sea bed (Foti et al., 2010) and the impact of weights or water jets on water (Watanabe and Tanaka, 2011). These pilot studies showed that, although more work is needed to use dinoflagellates as quantitative reporters of flow, the approach shows great promise for directly measuring theoretically modelled flow fields. For example, Latz and Rohr (2013) recently took advantage of high speed imaging of bioluminescent dinoflagellate cells to define the flow fields within bathyphotometers that differ in shape and stimulation strategy. These measurements are key for determining the optimal instrument design for accurate bioluminescence measurements in the ocean.

In situ measurements and field ecology applications

A variety of instrumentation currently exists for making *in situ* measurements of planktonic flow stimulated bioluminescence. Instruments, known as bathyphotometers, generally comprise of a light-tight chamber containing a form of light sensor, such as a photo-diode or photomultiplier tube, into which water is drawn at a known rate after being agitated to induce a bioluminescent response. A lack of standardization across instruments including differences in agitation methods and measurement approach (sampling frequency, calibration and processing) often makes the comparison of measurements from different studies difficulty (Bivens et al.s, 2002). However, a number of general patterns in the distribution of bioluminescence are known. Increased bioluminescence often coincides with areas of upwelling, such as that along the West African coast (Piontkovski et al., 1997), and at frontal zones (Lieberman et al., 1987; Cussatlegras et al., 2001). On a basin scale across the Atlantic, strong similarities have been found between the distribution of bioluminescence and chlorophyll-a (Haddock et al., 2010). The distribution of bioluminescence also varies with depth. Bioluminescence is typically highest within the mixed layer and decreases below the thermocline (Neilson et al., 1995) although exact profiles may change depending on location, water mass and time of year (Cussatlegras et al., 2001).

In situ measurement of bioluminescence has a number of applications for assessing the wider ocean environment. Specifically, it can be used to assess and monitor the structure and variability of marine ecosystems over a range of spatial and temporal scales (Kushnir et al., 1997). The spatial heterogeneity of bioluminescence can also be related to fronts, eddies and current meanders and consequently can be used to indicate the distribution of plankton biomass in relation to physical processes (Herring and Widder, 2001; Moline et al., 2007). Previous studies have used horizontal and vertical profiles of bioluminescence to identify internal wave motions (Kushnir et al., 1997), assess the small scale structure of plankton patchiness (Tokarev et al., 1998, 1999b) and the physiological state of planktonic ecosystems (Tokarev et al., 1999a). Bioluminescence has even been used to determine the effect of different seamount geomorphologies on the structure and distribution of the overlying pelagic communities (Tokarev et al., 2003).

Intensified cameras can be utilised to capture light stimulated from dinoflagellates as fish or other organisms move through the water (Clement et al., 2012). This allows the detection of fish schools at night from aircraft and their identification via the species specific patterns in bioluminescence they create (Squire and Krumboltz, 1981). Such methods have been used to evaluate stocks of anchovy, sardine, and tuna among other species across different regions including off the coast of Mexico and Chile (Squire and Krumboltz, 1981). Bioluminescence emissions can also be used in the study of larger animals. For example, video imaging of light emissions from dinoflagellate has previously been employed to study the flow field around dolphins as they move through the water (Rohr et al., 1998).

One growing application of measuring planktonic bioluminescence is for the detection and monitoring of Harmful Algal blooms (HAB) in near shore waters. Dinoflagellates are estimated to account for approximately 75% of HAB species (Smayda, 1997). Many bioluminescent dinoflagellates produce harmful toxins, such as *Alexandrium tamarense,* or red tides, such as *Noctilluca scintillans,* which can have a disastrous impact on local environments and aquaculture (Anderson et al., 2012). Increased bioluminescence has been observed to coincide with red tides in the Southern Sea of Korea, even though the species responsible for the red-tide was not bioluminescent (Kim et al., 2006). Time-series monitoring within the Finish Archipelago Sea has also identified that bioluminescence can be used to detect the onset of toxic blooms of *Alexandrium ostenfeldii* and study their spatiotemporal dynamics (Le Tortorec et al., 2013). Bioluminescence has the potential to provide a rapid biological marker for the presence of problem dinoflagellate species and could be used, in future, as part of an early warning system for harmful algal blooms in coastal regions.

References

Abrahams, M.V. and Townsend, L.D. (1993). Bioluminescence in dinoflagellates: A test of the burgular alarm hypothesis. Ecology. *74*, 258-260.

Alpermann, T.J., Tillmann, U., Beszteri, B, Cembella, A.D. and John, U. (2010). Phenotypic variation and genotypic diversity in a planktonic population of the toxigenic marine dinoflagellate *Alexandrium tamarense* (Dinophyceae). J. Phycol. *46*, 18-32.

Akimoto, H., Wu, C., Kinumi, T. and Ohmiya, Y. (2004). Biological rhythmicity in expressed proteins of the marine dinoflagellate *Lingulodinium polyedrum* demonstrated by chronological proteomics. Biochem. Biophys. Res. Commun. *315*, 306-312.

Anderson, D.M., Alpermann, T.J., Cembella, A.D., Collos, Y., Masseret, E. and Montresor, M. (2012). The globally distributed genus *Alexandrium*: Multifaceted roles in marine ecosystems and impacts on human health. Harmful Algae. *14*, 10-35.

Barcelos e Ramos, J., Muller, M.N. and Riebesell, U. (2010). Short-term response of the coccolithophore *Emiliania huxleyi* to an abrupt change in seawater carbon dioxide concentrations, Biogeosciences. *7*, 177-186.

Batchelder, H.P., Swift, E. and Keuren, J.R. (1992). Diel patterns of planktonic bioluminescence in the northern Sargasso Sea. Mar. Biol. *113*, 329-339.

Biggley, W.H., Swift, E., Buchanan, R.J. and Seliger, H.H. (1969). Stimulable and spontaneous bioluminescence in the marine dinoflagellates *Pyrodinium bahamense*, *Gonyaulax polyedra* and *Pyrocystis lunula*. J. Gen. Physiol. *54*, 96-122.

Bivens, J.R., Geiger, M.L. and Bird, J.L. (2002). Interpreting bathyphotometer signals., Oceans'02 MTS/IEEE, 3.

Brading, P., Warner, M.E., Davey, P., Smith, D.J., Achterberg, E.P. and Suggett, D.J. (2011). Differential effects of ocean acidification on growth and photosynthesis among phylotypes of *Symbiodinium* (Dinophyceae). Limnol. Oceanogr. *56*, 927–938.

Burkenroad, M.D. (1943). A possible function of bioluminescence. J. Mar. Res. *5*, 161-164.

Buskey, E., Mills, L. and Swift, E. (1983). The effects of dinoflagellate bioluminescence on the swimming behavior of a marine copepod. Limnol. Oceanogr. *28*, 575-579.

Buskey, E.J. and Swift, E. (1985). Behavioral responses of oceanic zooplankton to simulated bioluminescence. Biol. Bull. *168*, 263-275.

Buskey, E.J., Strom, S. and Coulter, C. (1992). Biolumiscence of heterotrophic dinoflagellates from Texas coastal waters. J. Exp. Mar. Biol. Ecol. *159*, 37-49.

Buskey, E.J., Coulter, C.J. and Brown, S.L. (1994). Feeding, growth and bioluminescence of the heterotrophic dinoflagellate *Protoperidinium huberi*. Mar. Biol. *121*, 373-380.

Caldeira, K., Wickett, M.E. (2003). Anthropogenic carbon and ocean pH. Nature *425*, 365-365.

Chen, A.K., Latz, M.I., Sobolewski, P. and Frangos, J.A. (2007). Evidence for the role of G-proteins in flow stimulation of dinoflagellate bioluminescence. Am. J. Physiol. Regul. Integr. Comp. Physiol. *292*, R2020-R2027.

Clement, P.M., Fjälling, A., Jaffe, J.S., Jones, E.G., Lundgren, B., Macaulay, G.J. and McElderry, H. (2012). Fishery applications of optical technologies. (Copenhagen, Denmark: International Council for the Exploration of the Sea).

Colepicolo, P., Roenneberg, T., Morse, D., Taylor, W.R. and Hastings, J.W. (1993). Circadian regulation of bioluminescence in the dinoflagellate *Pyrocystis lunula*. J. Phycol. *29*, 173-179.

Cussatlegras, A.-S., Geistdoerfer, P. and Prieur, L. (2001). Planktonic bioluminescence measurements in the frontal zone of Almeria–Oran (Mediterranean Sea). Oceanol. Acta. *24*, 239-250.

Dason, J.S., Huertas, I.E. and Colman, B. (2004). Source of inorganic carbon for photosynthesis in two marine dinoflagellates. J. Phycol. *40*, 285–92.

DeSa, R. and Hastings, J.W. (1968). The characterization of scintillons. Bioluminescent particles from the marine dinoflagellate, *Gonyaulax polyedra*. J. Gen. Physiol. *51*, 105-122.

Dunlap, J.C. and Hastings, J.W. (1981). The biological clock in *Gonyaulax* controls luciferase activity by regulating turnover. J. Biol. Chem. *256*, 10509-10518.

Erdner, D.L. and Anderson, D.M. (2006). Global transcriptional profiling of the toxic dinoflagellate *Alexandrium fundyense* using massively parallel signature sequencing. BMC Genomics. *7*, 88.

Esaias, W.E. and Curl, H.C. (1972). Effect of dinoflagellate bioluminescence on copepod ingestion rates. Limnol. Oceanogr. *17*, 901-905.

Esaias, W.E., Curl, H.C. and Seliger, H.H. (1973). Action spectrum for a low intensity, rapid photoinhibition of mechanically stimulable bioluminescence in the marine dinoflagellates *Gonyaulax catenella*, *G. acatenella*, and *G. tamarensis*. J. Cell. Physiol. *82*, 363-372.

Fleisher, K.J. and Case, J.F. (1995). Cephalopod predation facilitated by dinoflagellate luminescence. Biol. Bull. *189*, 263-271.

Fogel, M. and Hastings, J.W. (1972). Bioluminescence - Mechanism and mode of control of scintillon activity. Proc. Natl. Acad. Sci. U.S.A. *69*, 690-693.

Foti, E., Faraci, C., Foti, R. and Bonanno, G. (2010). On the use of bioluminescence for estimating shear stresses over a rippled seabed. Mecc. *45*, 881-895.

Fritz, L., Morse, D. and Hastings, J.W. (1990). The circadian bioluminescence rhythm of *Gonyaulax* is related to daily variations in the number of light-emitting organelles. J. Cell Sci. *95*, 321-328.

Frommlet, J.C. and Iglesias-Rodríguez, M.D. (2008). Microsatellite genotyping of single cells of the dinoflagellate species *Lingulodinium polyedrum* (Dinophyceae): a novel approach for marine microbial population genetic studies. J. Phycol. *44*, 1116-1125.

Greenblatt, P.R., Feng, D.F., Zirino, A. and Losee, J.R. (1984). Observations of planktonic bioluminescence in the euphotic zone of the California Current. Mar. Biol. *84*, 75-82.

Hinga, K.R. (1992). Co-occurrence of dinoflagellate blooms and high pH in marine enclosures. Mar. Ecol. Prog. Ser. *86*, 187-192.

Iglesias-Rodriguez, M.D., Schofield, O.M., Batley, J., Medlin, L.K., and Hayes, P.K. (2006). Intraspecific genetic diversity in the marine coccolithophorid *Emiliania huxleyi* (Prymnesiophyceae): the use of microsatellite analysis in marine phytoplankton population studies. J. Phycol. *42*, 526–536.

Iglesias-Rodriguez, M.D. (2013). Ocean Acidification. In Earth System Monitoring, Encyclopedia 269 of Sustainability Science and Technology, Orcutt, J., ed. (New York, U. S. A.: Springer Science).

Hackett, J.D., Scheetz, T.E., Yoon, H.S., Soares, M.B., Bonaldo, M.F., Casavant, T.L. and Bhattacharya, D. (2005). Insights into a dinoflagellate genome through expressed sequence tag analysis. BMC Genomics *6*, 80.

Haddock, S.H.D., Moline, M.A. and Case, J.F. (2010). Bioluminescence in the sea. Ann. Rev. Mar. Sci. *2*, 443-493.

Hamman, J.P., Biggley, W.H. and Seliger, H.H. (1981a). Photoinhibition of stimulable bioluminescence in marine dinoflagellates. Photochem. Photobiol. *33*, 909-914.

Hamman, J.P., Biggley, W.H. and Seliger, H.H. (1981b). Action spectrum for the photoinhibition of bioluminescence in the marine dinoflagellate *Dissodinium lunula*. Photochem. Photobiol. *33*, 741-747.

Hamman, J.P. and Seliger, H.H. (1982). The chemical mimicking of the mechanical stimulation, photoinhibition, and recovery from photoinhibition of bioluminescence in the marine dinoflagellate, *Gonyaulax polyedra*. J. Cell. Physiol. *111*, 315-319.

Harvey, E.N. (1957). A History of Luminescence from the Earliest Times Until 1900: From the Earliest Times Until 1900 (New York, U. S. A.: Dover Publications).

Hastings, J. (2013). Circadian Rhythms in Dinoflagellates: What Is the Purpose of Synthesis and Destruction of Proteins? Microorganisms. *1*, 26-32.

Hastings, J.W. (2007). The *Gonyaulax* clock at 50: translational control of circadian expression. Cold Spring Harbor Symp. Quant. Biol. *72*, 141-144.

Hastings, J.W. (2009). Dinoflagellate luminescence and its circadian regulation. Online Resource, available: http://photobiology.info/Hastings.html, accessed 06 June 2013.

Heimann, K., Klerks, P.L. and Hasenstein, K.H. (2009). Involvement of actin and microtubules in regulation of bioluminescence and translocation of chloroplasts in the dinoflagellate *Pyrocystis lunula*. Bot. Mar. *52*, 170-177.

Herring, P.J. and Widder, E.A. (2001). Bioluminescence in the plankton and nekton. In Steele, J.H., Thorpe, S. and Turekian, K., eds. Encyclopedia of Ocean Science. (San Diego, U. S. A. Academic Press), pp. 308-317.

Jaeckisch, N., Yang, I., Wohlrab, S., Glöckner, G., Kroymann, J., Vogel, H., Cembella, A. and John, U. (2011). Comparative genomic and transcriptomic characterization of the toxigenic marine dinoflagellate *Alexandrium ostenfeldii*. PLoS ONE. *6*, e28012.

Jin, K., Klima, J.C., Deane, G., Dale Stokes, M. and Latz, M.I. (2013). Pharmacological investigation of the bioluminescence signaling pathway of the dinoflagellate *Lingulodinium polyedrum*: evidence for the role of stretch-activated ion channels. J. Phycol. *49*, 733-745.

Johnson, C.H., Roeber, J.F. and Hastings, J.W. (1984). Circadian changes in enzyme concentration account for rhythm of enzyme activity in *Gonyaulax*. Science. *223*, 1428-1430.

Johnson, C.H., Inoue, S., Flint, A. and Hastings, J.W. (1985). Compartmentalization of algal bioluminescence - autofluorescence of bioluminescent particles in the dinoflagellate *Gonyaulax* as studied with image-intensified video microscopy and flow cytometry. J. Cell Biol. *100*, 1435-1446.

Kato, M., Chiba, T., Li, M. and Hanyu, Y. (2011). Bioluminescence assay for detecting cell surface membrane protein expression. Assay Drug Dev. Technol. *9*, 31-39.

Kelly, M.G. and Katona, S. (1966). An endogenous diurnal rhythm of bioluminescence in a natural population of dinoflagellates Biol Bull. *131*, 115-126.

Kelly, M.G. (1968). The occurrence of dinoflagellate luminescence at Woods Hole. Biol. Bull. *135*, 279-295.

Kim, G., Lee, Y.-W., Joung, D.-J., Kim, K.-R. and Kim, K. (2006). Real-time monitoring of nutrient concentrations and red-tide outbreaks in the southern sea of Korea. Geophys. Res. Lett. *33*, L13607.

Knaust, R., Urbig, T., Li, L.M., Taylor, W. and Hastings, J.W. (1998). The circadian rhythm of bioluminescence in *Pyrocystis* is not due to differences in the amount of luciferase: A comparative study of three bioluminescent marine dinoflagellates. J. Phycol. *34*, 167-172.

Krasnow, R., Dunlap, J., Taylor, W., Hastings, J.W., Vetterling, W. and Gooch, V. (1980). Circadian spontaneous bioluminescent glow and flashing of *Gonyaulax polyedra*. J. Comp. Physiol., B. *138*, 19-26.

Kushnir, V.M., Tokarev, Y.N., Williams, R., Piontkovski, S.A. and Evstigneev, P.V. (1997). Spatial heterogeneity of the bioluminescence field of the tropical Atlantic Ocean and its relationship with internal waves. Mar. Ecol. Prog. Ser. *160*, 1-11.

Langer, G., Nehrke, G., Probert, I., Ly, J. and Ziveri, P. (2009). Strain-specific responses of *Emiliania huxleyi* to changing seawater carbonate chemistry. Biogeosciences *6*, 2637–2646.

Lapointe, M. and Morse, D. (2008). Reassessing the role of a 3'-UTR-binding translational inhibitor in regulation of circadian bioluminescence rhythm in the dinoflagellate *Gonyaulax*. Biol. Chem. *389*, 13-19.

Lapointe, M., MacKenzie, T.D.B. and Morse, D. (2008). An external delta-carbonic anhydrase in a free-living marine dinoflagellate may circumvent diffusion-limited carbon acquisition. Plant Physiol. *147*, 1427–36.

Lapota, D. and Losee, J.R. (1984). Observations of bioluminescence in marine plankton from the Sea of Cortez. J. Exp. Mar. Biol. Ecol. *77*, 209-239.

Lapota, D., Rosenberger, D.E. and Lieberman, S.H. (1992a). Planktonic bioluminescence in the pack ice and the marginal ice zone of the Beaufort Sea. Mar. Biol. *112*, 665-675.

Lapota, D., Young, D.K., Bernstein, S.A., Geiger, M.L., Huddell, H.D. and Case, J.F. (1992b). Diel bioluminescence in heterotrophic and photosynthetic marine dinoflagellates in an Arctic fjord. J. Mar. Biol. Assoc. U.K. *72*, 733-744.

Lapota, D., Osorio, A.R., Liao, C. and Bjorndal, B. (2007). The use of bioluminescent dinoflagellates as an environmental risk assessment tool. Mar. Pollut. Bull. *54*, 1857-1867.

Latz, M.I. and Lee, A.O. (1995). Spontaneous and stimulated bioluminescence of the dinoflagellate *Ceratocorys horrida* (Peridiniales). J. Phycol. *31*, 120-132.

Latz, M.I. and Jeong, H.J. (1996). Effect of red tide dinoflagellate diet and cannibalism on the bioluminescence of the heterotrophic dinoflagellates *Protoperidinium* spp. Mar. Ecol. Prog. Ser. *132*, 275-285.

Latz, M.I., Bovard, M., VanDelinder, V., Segre, E., Rohr, J. and Groisman, A. (2008). Bioluminescent response of individual dinoflagellate cells to hydrodynamic stress measured with millisecond resolution in a microfluidic device. J. Exp. Biol. *211*, 2865-2875.

Latz, M.I. and Rohr, J. (2013). Bathyphotometer bioluminescence potential measurements: A framework for characterizing flow agitators and predicting flow-stimulated bioluminescence intensity. Cont. Shelf Res. *61–62*: 71-84.

Le Tortorec, A.H., Hakanen, P., Kremp, A., Olsson, J., Suikkanen, S. and Simis, S.G.H. (2013). Stimulated bioluminescence as an early indicator of bloom development of the toxic dinoflagellate *Alexandrium ostenfeldii*. J. Plankton Res. bt116.

Lee, D.H., Mittag, M., Sczekan, S., Morse, D. and Hastings, J.W. (1993). Molecular cloning and genomic organization of a gene for luciferin-binding protein from the dinoflagellate *Gonyaulax polyedra*. J. Biol. Chem. *268*: 8842-8850.

Li, L., Hong, R. and Hastings, J.W. (1997). Three functional luciferase domains in a single polypeptide chain. Proc. Natl. Acad. Sci. U.S.A. *94*, 8954-8958.

Li, L. and Hastings, J.W. (1998). The structure and organization of the luciferase gene in the photosynthetic dinoflagellate *Gonyaulax polyedra*. Plant Mol. Biol. *36*, 275-284.

Li, L., Liu, L., Hong, R., Robertson, D. and Hastings, J.W. (2001). N-terminal intramolecularly conserved histidines of three domains in *Gonylaulax* luciferase are responsible for loss of activity in the alkaline region. Biochemistry. *40*, 1844-1849.

Li, L., Higaki, K., Ninomiya, H., Luan, Z., Iida, M., Ogawa, S., Suzuki, Y., Ohno, K. and Nanba, E. (2010). Chemical chaperone therapy: Luciferase assay for screening of β-galactosidase mutations. Mol. Genet. Metab. *101*, 364-369.

Li, Y.Q., Swift, E. and Buskey, E.J. (1996). Photoinhibition of mechanically stimulable bioluminescence in the heterotrophic dinoflagellate *Protoperidinium depressum* (Pyrrophyta). J. Phycol. *32*, 974-982.

Lieberman, S.H., Lapota, D., Losee, J.R. and Zirino, A. (1987). Planktonic Bioluminescence in the Surface Waters of the Gulf of California. Biol. Oceanogr. *4*, 25-46.

Liu, L.Y., Wilson, T. and Hastings, J.W. (2004). Molecular evolution of dinoflagellate luciferases, enzymes with three catalytic domains in a single polypeptide. Proc. Natl. Acad. Sci. U.S.A. *101*, 16555-16560.

Liu, L.Y. and Hastings, J.W. (2007). Two different domains of the luciferase gene in the heterotrophic dinoflagellate *Noctiluca scintillans* occur as two separate genes in photosynthetic species. Proc. Natl. Acad. Sci. U.S.A. *104*, 696-701.

Lynch, R.V. (1978). The occurence and distribution of surface bioluminescence in the oceans during 1966 through 1977 (Washington DC, U.S.A.: Naval Research Lab).

Machabée, S., Wall, L. and Morse, D. (1994). Expression and genomic organization of a dinoflagellate gene family. Plant Mol. Biol. *25*, 23-31.

Maldonado, E.M. and Latz, M.I. (2007). Shear-stress dependence of dinoflagellate bioluminescence. Biol. Bull. *212*, 242-249.

Marcinko, C.L.J., Allen, J.T., Poulton, A.J., Painter, S.C. and Martin, A.P. (2013). Diurnal variations of dinoflagellate bioluminescence within the open-ocean north-east Atlantic. J. Plankton Res. *35*, 177-190.

Mesinger, A.F. and Case, J.F. (1992). Dinoflagellate luminescence increases susceptibility of zooplankton to teleost predation. Mar. Biol. *112*, 207-210.

Mittag, M., Lee, D.H. and Hastings, J.W. (1994). Circadian expression of the luciferin-binding protein correlates with the binding of a protein to the 3' untranslated region of its mRNA. Proc. Natl. Acad. Sci. U.S.A. *91*, 5257-5261.

Mittag, M., Li, L. and Hastings, J.W. (1998). The mRNA level of the circadian regulated *Gonyaulax* luciferase remains constant over the cycle. Chronobiol. Int. *1998*, 93-98.

Moline, M.A., Oliver, M.J., Mobley, C.D., Sundman, L., Bensky, T., Bergmann, T., Bissett, W.P., Case, J., Raymond, E.H. and Schofield, O.M.E. (2007). Bioluminescence in a complex coastal environment: 1. Temporal dynamics of nighttime water-leaving radiance. J.Geophys.Res.-Oceans 112.

Moline, M.A., Blackwell, S.M., Case, J.F., Haddock, S.H.D., Herren, C.M., Orrico, C.M. and Terrill, E. (2009). Bioluminescence to reveal structure and interaction of coastal planktonic communities. Deep-Sea Res. Pt II. *56*, 232-245.

Morse, D., Milos, P.M., Roux, E. and Hastings, J.W. (1989). Circadian regulation of bioluminescence in *Gonyaulax* involves translational control. Proc. Natl. Acad. Sci. U.S.A. *86*, 172-176.

Nakamura, H., Kishi, Y., Shimomura, O., Morse, D. and Hastings, J.W. (1989). Structure of dinoflagellate luciferin and its enzymic and nonenzymic air-oxidation products. J. Am. Chem. Soc. *111*, 7607-7611.

Neilson, D.J., Latz, M.I. and Case, J.F. (1995). Temporal variability in the vertical structure of bioluminescence in the North Atlantic Ocean. J. Geophys. Res. *100*, 6591-6603.

Nicolas, M.T., Sweeney, B.M. and Hastings, J.W. (1987). The ultrastructural localization of luciferase in three bioluminescent dinoflagellates, two species of *Pyrocystis*, and *Noctiluca*, using anti-luciferase and immunogold labelling. J. Cell Sci. *87*, 189-196.

Okamoto, O.K., Liu, L., Robertson, D.L. and Hastings, J.W. (2001). Members of a dinoflagellate luciferase gene family differ in synonymous substitution rates. Biochemistry. *40*, 15862-15868.

Reinfelder, J.R. (2011). Carbon concentrating mechanisms in eukaryotic marine phytoplankton, Annu. Rev. Mar. Sci. *3*: 291–315.

Piontkovski, S.A., Yu, N.T., Bitukov, E.P., Williams, R. and Kiefer, D.Ý. (1997). The bioluminescent field of the Atlantic Ocean. Mar. Ecol. Prog. Ser. *156*, 33-41.

Poupin, J., Cussatlegras, A.S. and Geistdoerfer, P. (1999). Plancton Marin Bioluminescent: nventaire documenté des espèces et bilan des formes les plus communes de la mer d'Iroise. (Brest, France: Diss. Ecole Navale). (In French).

Rohr, J., Latz, M.I., Fallon, S., Nauen, J.C. and Hendricks, E. (1998). Experimental approaches towards interpreting dolphin-stimulated bioluminescence. J. Exp. Biol. *201*, 1447-1460.

Rohr, J., Hyman, M., Fallon, S. and Latz, M.I. (2002). Bioluminescence flow visualization in the ocean: an initial strategy based on laboratory experiments. Deep-Sea Res. Pt I *49*, 2009-2033.

Roy, S., Letourneau, L. and Morse, D. (2014). Cold-Induced Cysts of the Photosynthetic Dinoflagellate Lingulodinium polyedrum Have an Arrested Circadian Bioluminescence Rhythm and Lower Levels of Protein Phosphorylation. Plant Physiol. *164*: 966-977.

Rynearson, T.A., Newton, J.A. and Armbrust, E.V. (2006). Spring bloom development, genetic variation, and population succession in the planktonic diatom *Ditylum brightwellii*. Limnol. Oceanogr. *51*, 1249-1261.

Schmitter, R.E., Njus, D., Sulzman, F.M., Gooch, V.D. and Hastings, J.W. (1976). Dinoflagellate bioluminescence - Comparative study of *in vitro* components. J. Cell. Physiol. *87*, 123-134.

Schultz, L.W., Liu, L., Cegielski, M. and Hastings, J.W. (2005). Crystal structure of a pH-regulated luciferase catalyzing the bioluminescent oxidation of an open tetrapyrrole. Proc. Natl. Acad. Sci. U.S.A. *102*, 1378-1383.

Seliger, H.H., Fastie, W.G., Taylor, W.R. and McElroy, W.D. (1962). Bioluminescence of Marine Dinoflagellates: I. An underwater photometer for day and night measurements. J. Gen. Physiol. *45*, 1003-1017.

Seo, K.S. and Fritz, L. (2000). Cell ultrastructural changes correlate with circadian rhythms in *Pyrocystis lunula* (Pyrrophyta). J. Phycol. *36*, 351-358.

Shimomura, O. (2006). Bioluminescence: chemical principles and methods (Singapore: World Scientific Publications).

Shulman, I., Moline, M.A., Penta, B., Anderson, S., Oliver, M. and Haddock, S.H.D. (2011). Observed and modeled bio-optical, bioluminescent, and physical properties during a coastal upwelling event in Monterey Bay, California. J. Geophys. Res. *116*, C01018.

Smayda, T.J. (1997). Harmful algal blooms: their ecophysiology and general relevance to phytoplankton blooms in the sea. Limnol. Oceanogr. *42*, 1137-1153.

Smith, S.M.E., Morgan, D., Musset, B., Cherny, V.V., Place, A.R., Hastings, J.W. and DeCoursey, T.E. (2011). Voltage-gated proton channel in a dinoflagellate. Proc. Natl. Acad. Sci. U.S.A. *108*, 18162-18167.

Soli, G. (1966). Bioluminescent cycle of photosynthetic dinoflagellates. Limnol. Oceanogr. *11*, 355-363.

Squire, J.L. and Krumboltz, H. (1981). Profiling pelagic fish schools using airborne optical lasers and other remote sensing techniques. Mar. Technol. Soc. J. *15*, 27-31.

Sullivan, J.M. and Swift, E. (1994). Photoinhibition of mechanically stimulable bioluminescence in the autotrophic dinoflagellate *Ceratium fusus* (Pyrrophyta). J. Phycol. *30*, 627-633.

Sullivan, J.M. and Swift, E. (1995). Photoenhancement of bioluminescence capacity in natural and laboratory populations of the autotrophic dinoflagellate *Ceratium fusus* (Ehrenb.) Dujardin. J. Geophys. Res. *100*, 6565-6574.

Suzuki-Ogoh, C., Wu, C. and Ohmiya, Y. (2008). C-terminal region of the active domain enhances enzymatic activity in dinoflagellate luciferase. Photochem. Photobiol. Sci. *7*, 208-211.

Suzuki, C., Nakajima, Y., Akimoto, H., Wu, C. and Ohmiya, Y. (2005). A new additional reporter enzyme, dinoflagellate luciferase, for monitoring of gene expression in mammalian cells. Gene. *344*, 61-66.

Sweeney, B.M., Haxo, F.T. and Hastings, J.W. (1959). Action spectra for two effects of light on luminescence in *Gonyaulax polyedra*. J. Gen. Physiol. *43*, 285-299.

Sweeney, B.M. 1981. Variations of the bioluminescence per cell in dinoflagellates. In Nealson, K.H., , ed. Bioluminescence Current Perspectives. (New York, U.S.A.: Burgess Publishing), pp. 90-94.

Swift, E. and Meunier, V. (1976). Effects of light intensity on division rate, stimulable bioluminescence and cell size of oceanic dinoflagellates *Dissodinium lunula*, *Pyrocystis fusiformis* and *Pyrocystis noctiluca*. J. Phycol. *12*, 14-22.

Swift, E., Sullivan, J.M., Batchelder, H.P., Van Keuren, J., Vaillancourt, R.D. and Bidigare, R.R. (1995). Bioluminescent organisms and bioluminescence measurements in the

North Atlantic Ocean near latitude 59.5 N, longitude 21 W. J. Geophys. Res. *100*, 6527-6547.
Tett, P.B. (1971). The relation between dinoflagellates and the bioluminescence of sea water. J. Mar. Biol. Assoc. U.K. *51*, 183-206.
Tett, P.B. and Kelly, M.G. (1973). Marine Bioluminescence. Oceanogr. Mar. Biol. Annu. Rev. *11*, 89-173.
Tokarev, Y., Bitukov, E.P., Williams, R., Vasilenko, V.I., Piontkovski, S.A. and Sokolov, B.G. 1999a. The bioluminescence field as an indicator of the spatial structure and physiological state of the planktonic community at the mediterranean sea basin. In Malanotte-Rizzoli, P. and Eremeev, V., eds. The Eastern Mediterranean as a Laboratory Basin for the Assessment of Contrasting Ecosystems. (Rotterdam, Netherlands: Springer), pp. 407-416.
Tokarev, Y.N., Williams, R. and Piontkovski, S.A. (1998). Small-scale plankton patchiness in the Black Sea euphotic layer. In Baden, S., et al., eds. Recruitment, Colonization and Physical-Chemical Forcing in Marine Biological Systems. (Rotterdam, Netherlands: Springer), pp. 363-367.
Tokarev, Y.N., Williams, R. and Piontkovski, S.A. (1999b). Identification of small-scale structure of plankton communities of the Black and Ionian Seas by their bioluminescence characteristics. Hydrobiologia. *393*, 163-167.
Tokarev, Y.N., Bityukov, E. and Williams, R. (2003). Influence of seamounts at the Atlantic Ocean on modification of the bioluminescence and plankton characteristics. Mar. Ecol. J. *2*, 46-58.
Topalov, G. and Kishi, Y. (2001). Chlorophyll catabolism leading to the skeleton of dinoflagellate and krill luciferins: hypothesis and model studies. Angew. Chem. *40*, 3892-3894.
Uribe, P., Fuentes, D., Valdés, J., Shmaryahu, A., Zúñiga, A., Holmes, D. and Valenzuela, P.D.T. (2008). Preparation and analysis of an expressed sequence tag library from the toxic dinoflagellate *Alexandrium catenella*. Mar. Biotechnol. *10*, 692-700.
Utyushev, R.N., Levin, L.A. and Gitelson, J.I. (1999). Diurnal rhythm of the bioluminescent field in the ocean epipelagic zone. Mar. Biol. *134*, 439-448.
Valiadi, M., Iglesias-Rodriguez, M.D. and Amorim, A. (2012). Distribution and genetic diversity of the luciferase gene within marine dinoflagellates. J. Phycol. *48*, 826-836.
Valiadi, M. and Iglesias-Rodriguez, M. (2013). Understanding bioluminescence in dinoflagellates - How far have we come? Microorganisms. *1*, 3-25.
Valiadi, M. and Iglesias-Rodriguez, M.D. (2014). Diversity of the luciferin binding protein gene in bioluminescent dinoflagellates – Insights from a new gene in *Noctiluca scintillans* and sequences from Gonyaulacoid genera. J. Eukaryot. Microbiol. *61*, 134-145.
Vanormelingen, P., Verleyen, E. and Vyverman, W. (2008). The diversity and distribution of diatoms: from cosmopolitanism to narrow endemism. Biodivers. Conserv. *17*, 393–405.
von Dassow, P. and Latz, M.I. (2002). The role of Ca^{2+} in stimulated bioluminescence of the dinoflagellate *Lingulodinium polyedrum*. J. Exp. Biol. *205*, 2971-2986.
Watanabe, Y. and Tanaka, Y. (2011). Bioluminescence-based imaging technique for pressure measurement in water. ExFl. *51*, 225-236.
White, H.H. (1979). Effects of dinoflagellate bioluminescence on the ingestion rates of herbivorous zooplankton. J. Exp. Mar. Biol. Ecol. *36*, 217-224.
Widder, E.A. (2001). Marine Bioluminescence. Biosc. Explain. *1*, 1-9.
Wilson, T. and Hastings, J. (2013). Bioluminescence: Living lights, lights for living. (Massachussets, U.S.A.: Harvard University Press).

Yentsch, C.S. and Laird, J.C. (1968). Seasonal sequences of bioluminescence and the occurrence of endogenous rhythm in oceanic waters of Woods Hole, Massachusetts. J. Mar. Res. *26*, 127-133.

Yoo, K.I. (1991). Population dynamics of dinoflagellate community in Masan Bay with a note on the impact of environmental parameters. Mar. Poll. Bull. *23*, 185–188.

Other Publications of Interest

Flores, E. and Herrero, A. (2014). The Cell Biology of Cyanobacteria. Caister Academic Press, U.K. ISBN: 978-1-908230-38-6

Herold, K.E. and Rasooly, A. (2009). Lab-on-a-Chip Technology: Fabrication and Microfluidics. Caister Academic Press, U.K. ISBN: 978-1-904455-46-2

Herold, K.E. and Rasooly, A. (2009) Lab-on-a-Chip Technology: Biomolecular Separation and Analysis. Caister Academic Press, U.K. ISBN: 978-1-904455-47-9

Wilkinson, M.G. (2015). Flow Cytometry in Microbiology: Technology and Applications. Caister Academic Press, U.K. ISBN: 978-1-910190-11-1

Xuefeng, L. (2014). Biofuels: From Microbes to Molecules. Caister Academic Press, U.K. ISBN: 978-1-908230-45-4

7

Microfluidics and *in situ* Sensors for Microalgae

Maria-Nefeli Tsaloglou[1,2*]

[1]Ocean and Earth Science and Institute for Life Sciences, University of Southampton, Southampton, Hampshire, SO17 1BJ, U.K. M.Tsaloglou@soton.ac.uk
[2]Chemistry and Chemical Biology, Harvard University, Cambridge MA02138, U.S.A. Tsaloglou@gmwgroup.harvard.edu
*Author for correspondence.

Contents

Abstract

Phytoplankton are important in global biogeochemistry since they produce the bulk of oxygen on Earth through photosynthesis. They form the base of the marine food web and are primary producers of organic carbon. Some species produce polyether toxins, thus forming harmful algal blooms that can have detrimental effects on local fauna and flora.

For these reasons, phytoplankton monitoring in the oceans is critical. It can provide insight to climate change, microbial ecology, biogeochemical cycles and health toxicity of recreational waters. Currently, the oceans are vastly under-sampled and *in situ* sensors and sensor networks can address this need for sampling. Miniaturised and microfluidic sensors are increasingly being used for phytoplankton monitoring with great promise for future applications. This chapter will provide an overview of traditional methods of phytoplankton analysis, introduce microfluidics, and review the state-of-the art in miniaturised and *in situ* sensors for algal detection in the oceans.

Introduction

Data collection in the oceans is typically done during organised cruises. These expeditions can be very expensive, labour-intensive and usually target pre-defined locations. As a consequence, the oceans are vastly under-sampled; only about 5 percent of the 1.3 billion cubic kilometres covering Earth have been explored until now (Schofield et al., 2013). Samples are collected in the field and then analysed on-board research vessels or preserved for *ex situ* studies in laboratories. However, not all phytoplankton can be preserved, which results in preservation bias.

Due to the massive volume of water that would need to be sampled for complete characterisation of global marine, microbial, taxonomic and functional diversity, and of microalgae in particular, the task is almost insurmountable using solely data/sample collection from research cruises. Remote sensing using ocean colour analysis of satellite images is a more cost-effective approach to obtain information on phytoplankton biomass, based on chlorophyll concentration (Blondeau-Patissier et al., 2014). Its disadvantages are that data cannot be obtained in cloudy weather, are limited to first optical depth, and spatial resolution can be low.

Autonomous Underwater Vehicles (AUVs) and other complementary platforms implement data collection for phytoplankton in the oceans by housing *in situ* and deployable sensors (Schofield et al., 2013). A sensor is defined as an integrated analytical device that does not require reagent addition and includes a recognition element for analyte detection, as known as a transducer (Figure 7.1). This can be a substance or a device that converts the sensing signal to a detectable output, quantitative or qualitative, such as electrical

Figure 7.1. A simple diagram of a sensor. The sample is prepared for analysis and the transducer senses a signal input. The signal is converted to a detectable ouput and then interpreted.

energy or light. A biosensor, in particular, has a biological transducer, which would be an enzyme, an antibody or whole cells.

Sensors for phytoplankton can be deployed while completely submerged and their components include sample processing, analysis and data storage. The majority of these platforms are not simply a sensor, as described by the definition, but large integrated systems with multiple sensing components. They can provide biological, physical, chemical or optical information by employing analytical tools, which are portable versions of the equipment used *ex situ* in the lab or on-board research ships for phytoplankton analysis.

The state-of-the-art for lab-on-chip technologies for algal detection using optical and impedance methods (Schaap et al., 2012a), as well as *in situ* sensors for phytoplankton (Erickson et al., 2011) and nutrient (Mills and Fones, 2012) analyses have been reviewed in 2012. In this chapter, we will review the most significant work in the prior art using all detection methods (biological, physical, chemical, including optical), and focus on advances in the field after 2012.

Traditional methods for analysis of microalgae

The most common methods of algal analysis and identification are optical and electron microscopy (Tomas, 1997). Cells are stained with fluorescence or immunological methods to assist visualisation. This process is time-consuming and requires specialised cytologists. Alternatively, computer-based microscopy image recognition analyses can be used to identify and count phytoplankton (Embleton et al., 2003) with an open source software, PlanktoVision, recently developed (Schulze et al., 2013). Light scattering, particle sizing and spectrophotometry are also used for initial size and population information.

Photosynthetic pigment identification can be used to distinguish between different phytoplankton species (Jeffrey et al., 2011). Non-fluorescent algal pigments are detected with High-Pressure Liquid Chromatography (HPLC) (Bidigare et al., 2005; Irigoien et al., 2004) and Liquid Chromatography Mass Spectrometry (LC-MS) (Airs and Garrido, 2011). Recent papers have used micro-Raman spectroscopy to map carotenoids in *Chlorella sorokiniana* and *Neochloris oleobundans* (Huang et al., 2010; Pilát et al., 2012).

Fluorometry is used for bulk characterisation of phytoplankton fluorescence. Flow cytometric analysis of phytoplankton yields quantitative data based on the auto-fluorescence of cells from their photosynthetic pigments, but also their size and light scattering properties (Sosik et al., 2010). It has been used as a shipboard tool since the 1980s (Olson et al., 1985). In particular, Fluorescence-Activated Cell Sorting (FACS) can separate a heterogeneous population of algae in taxonomic classes with respect to their fluorescent pigments. Some examples of pigments are (1) chlorophyll (found in all phytoplankton) that emits red fluorescence when excited at 488 nm; (2) phycoerythrine (more abundant in cyanobacteria/cryptophytes) that emits orange fluorescence when excited at 488 nm and (3) phycocyanine (present in cyanobacteria/cryptophytes) that emits red fluorescence when excited at 633 nm.

Native algal cells from the environmental samples obtained in the field can be isolated and either preserved for future analyses, or grown in the lab using microbiological methods. This allows for further investigations and has resulted in several culture collections available word-wide. In the United Kingdom, these are the Plymouth Culture Collection of Marine Microalgae (CCMM) in England, and the Culture Collection of Algae and Protozoa (CCAP) in Scotland. The US-based national collection is housed at the National Center for Marine Algae and Microbiota (CMAM), formerly known as the Provasoli-Guillard National Center for Culture of Marine Phytoplankton (CCMP).

Cells can then be further analysed using molecular biology methods (Zehr et al., 2009). Nucleic acids found in minute amounts in cells require amplification to detectable limits for further analyses by Nucleic Acid Amplification Technologies (NAAT). DNA can be amplified by Polymerase Chain Reaction (PCR) (Mullis et al., 1987) and RNA with reverse transcriptase-PCR (RT-PCR). More recent amplification alternatives, termed as isothermal, operate at lower temperatures than PCR and do not require thermo-cycling; their state-of-the-art has been recently reviewed (Yan et al., 2014). However, amplification of genes of interest by NAAT creates bias when studying microbial diversity. Molecular methods without DNA amplification for the investigation of mixed community phytoplankton samples include genome sequencing, termed as genomics, and metagenomics (Heidelberg and Heidelberg, 2005; Rynearson and Palenik, 2011). Metagenomics or community genomics, unlike genomics, do not rely on isolating individual species and apply Sanger or pyro-sequencing methods to environmental samples of entire communities. The United States Joint Genome Institute sponsored by the Department of Energy collects and catalogues algal genomic data. A special mention has to be made on J. Craig Venter's approach for building a metagenomic profile of the oceans during the global ocean sampling expedition on board the Sorcerer II sailing vessel (Rusch et al., 2007; Williamson et al., 2012). Phytoplankton proteomics (Jones et al., 2011) and metaproteomics (Siggins et al., 2012) are far less explored areas, as is lipidomics (Li et al., 2014) with only a handful of studies as of yet.

Some mainstream bench-top molecular tools have been further optimised specifically for microbial community analysis applications. Firstly, the PhyloChip is a microarray with 297 851 immobilized probes for 16S rRNA genes (DeSantis et al., 2007) on a glass chip, developed by the Lawrence Berkeley National Laboratory (U.S.A.). The 25-mer probes are complementary to 842 bacterial and archeal subfamilies and can provide with high precision community information from a single sample. The principle of operation is that DNA or DNA extracted from an environmental sample is first labelled with a fluorescent dye. The native fluorescent DNA is then allowed to hybridise with the probes and any unbound nucleic acids are washed off. The signal intensity corresponding to different probes is finally quantified. Secondly, the GeoChip follows the microarray design concept with 167 044 50-mer oligos covering 395 894 coding sequences from bacteria, achaea, fungi and viruses (He et al., 2007). Thirdly, the Biotrove Open Array system has been used extensively for environmental samples (van Doorn et al., 2009). It allows high throughput qPCR reactions in 33 nL volumes. A total of 3 072 reactions can be run in each plate and real-time fluorescence is monitored using SYBR or Taqman probes.

Microfluidics or lab-on-chip

Microfluidics is the science and technology of systems that process or manipulate small amounts of fluids using channels with dimensions from one to 1 000 μm (Whitesides, 2006). The field is largely interdisciplinary, combining aspects of engineering, physics, chemistry and biology. The concept of miniaturised total chemical analysis systems was presented in the early 90s by Andreas Manz (Manz et al., 1990), although some embryonic elements were introduced in the 1950s and later developed to the current inkjet printing technology. The early work of Manz *et al* on a miniaturised high-pressure liquid chromatography (HPLC) column microfluidic device (Manz et al., 1990) initiated the field that is now widely known as "lab-on-chip". Since then, a myriad of devices has been developed for applications varying from environmental analysis (Li and Lin, 2009), forensics (Horsman et al., 2007) and complex clinical samples (Crevillén et al., 2007).

A microfluidic platform can perform the same set of operations as larger fluidic alternatives for chemical and biochemical processes and reactions typically taking place in specialised laboratories (Mark et al., 2010). The advantages of microfluidic systems are; (i) ease of operation without the need of highly trained personnel; (ii) reduced reagent consumption, cost of manufacture and waste generation; (iii) low power usage; (iv) and significantly smaller size. Lab-on-chip devices also allow parallelisation of reactions and thus yielding faster results with better statistics. Systems can be modular and mimic *in vivo* conditions. Finally, single cell and molecule applications are feasible with microfluidics since they match well with the small size of systems.

Devices can be fabricated in polymer substrates (thermoplastics or elastomers) or in glass, both having the advantage of being optically clear. A large variety of approaches for particle separation, fluid manipulation, device fabrication and operation have been presented. A few examples will be highlighted hereon. Continuous flow is typically effected using pumps, mixers and valves. One example is the Quake pneumatic valve constructed from polydimethylsiloxane (PDMS) by soft-lithographic procedures (Hansen et al., 2002). It operates by restriction of a microfluidic channel using pressure and allows design of complicated chips with multiple fluidic paths. Alternative ways of fluid manipulation other than continuous flow are droplet microfluidics and digital microfluidics. Droplet microfluidics uses discreet mono-disperse droplets inside microfluidic channels formed by two immiscible fluids like oil and water (Teh et al., 2008). Digital microfluidics or electrowetting-on-dielectrics (EWOD) operates by applying a potential to an electrode coated with a hydrophobic insulator (Zeng and Wang, 2013). This modifies the contact angle of a drop of liquid lying on top of the electrode and allows mixing, dispensing and merging of very small droplets of water in oil (Choi et al., 2012). Finally, dielectrophoresis subjects particles to an AC field in order to induce a temporary dipole moment (Morgan et al., 1999). This allows accurate particle separation based on their polarisability.

Applications of microfluidics range from biological and chemical assays (Fair et al., 2007), chemical sensing (Wu and Gu, 2011) to optics (Tang et al., 2008), drug delivery (Khan et al., 2013), cell (Huang et al., 2002) and molecular (Kim et al., 2007) separation. As highlighted by George Whitesides in 2006 (Whitesides, 2006), advances in molecular analysis, biodefense, molecular biology and microelectronics are contributing to the field. The global market growth for microfluidic devices has been estimated to $5.7 billion by

2018 (Roussel, 2013), mostly driven by pharmaceutical companies for healthcare diagnostics.

There are two main examples of lab-on-chip devices that have been commercialised and are used for mainstream applications. Firstly, the Agilent Bioanalyzer is a microfluidic system, based on capillary electrophoresis, for analysis of nucleic acids and proteins. Secondly, the BioMark HD System from Fluidigm incorporates 9216 qPCR reactions on one single chip.

There is still large scope for innovative commercial devices since a complete sample-to-answer integrated system has yet to be developed. Many individual microfluidic components have been presented but major challenges still lie with sample preparation (Labuz and Takayama, 2014; Mariella, 2008) and system integration (Nge et al., 2013). Nevertheless, advances in the lab-on-chip field by 2014 have made the possibility of microfluidic devices becoming widely-used products a not too distant reality (Whitesides, 2014).

Deployable and *in situ* Sensors for Microalgae

Platforms and sensor networks

There is a large range of complementary platforms in addition to research ships and satellites for sampling and data acquisition on phytoplankton and microbial ecology (Shade et al., 2009). Stationary, fixed-point or Eulerian platforms are moorings, buoys and cabled observatories that offer high temporal resolution (Lampitt et al., 2010).They can offer large power capabilities and limited restriction on the size and shape of sensors that they can accommodate. However, they need to operate in large networks to provide spatial resolution (Adornato et al., 2010). Conversely, AUVs, gliders and Argo floats are unmanned mobile devices that operate autonomously, each type to a different extent. They can carry a variety of sensors and can be deployed for up to months at a time (Schofield et al., 2013). Their drawbacks are significant limitations on power and size of payload. Both stationary and mobile platforms suffer from biofouling due to long-term exposure to the marine environment (Delauney et al., 2010).

Sensors for morphology observations

As *in situ* alternatives to microscopy in a lab, the majority of sensors employ light scattering, spectrophotometry, fluorometry and imaging methods to analyse the size and shape of microalgae. Microfluidic systems for phytoplankton identification have been presented for impedance spectroscopy and optofluidic monolithic devices. These microsystems have not as of yet been deployed in the field, but show potential to integration to AUVs because of their small size and power consumption. An excellent review on optical tools available for ocean monitoring was published in 2009 (Moore et al., 2009).

Light scattering

Light scattering sensors are often referred to as turbidity sensors. They provide with information of particle size, composition and distribution of populations. They operate on the principle that light scatters in all directions when it hits a particle. Forward scattering correlates to the particle size and side scattering is indicative of particle shape. A large range of commercial turbidity sensors are now available (Moore et al., 2009) and established configuration standards like ISO 7027 exist.

The Bellouard group in the Netherlands developed an optofluidic microdevice for algal identification (Schaap et al., 2012a). A glass device, comprising of a straight channel and a curved waveguide, was etched with regions of higher refractive index using a femtosecond laser (Davis et al., 1996). A 1 550 nm laser excited algae passing through the microfluidic channel, producing fluctuations in optical intensity measured by a photodetector. Images of the individual algae were also captured by a video camera implemented on the device. The acquired wave-functions identified up to five separate classes of algae in a spiked sample with an accuracy of 78 percent (Schaap et al., 2012b). Environmental samples would be harder to discern on this system due to detritus present. Dilutions would be needed until only one cell was present in the detection zone at any one time. However, the monolithic nature of the device shows great promise for future adaption to a deployable sensor.

Spectrophotometry

Spectrophotometry measures the light absorbed by a sample as function of wavelength after illumination with broadband radiation or white light. An *in situ* spectrophotometer, the "*Brevebuster*", is available commercially for identification of *Karenia brevis* blooms (Robbins et al., 2006). The sensor uses a tungsten/deuterium light for illumination and a fiber optic spectrophotometer for detection. It has been used on moorings, buoys and ships (Kirkpatrick et al., 2011). The National Oceanic and Atmospheric Administration (NOAA) deploys a network of these "*Brevebuster*" systems on fixed platforms in Tampa Bay (Florida, U.S.A.) for automated detection of red tides

Fluorometry and bioluminescence

Fluorescence is the process of light emission by a sample when excited by a specific stimulus. It provides with information that allows distinguishing algal cells from debris and other particles in seawater. Identification of major phytoplankton groups is also feasible based on the light absorbed by their predominant photosynthetic pigments and subsequently emitted as fluorescent light. A variety of commercial instruments are available and one indicative example for each type of fluorometers will be illustrated in the following section. The list is by no means exhaustive and the author holds no commercial relationships with the mentioned companies.

Single channel fluorometers measure one single wavelength and are calibrated for detection of chlorophyll, phycocyanine (or other pigments), and dissolved organic matter. The Amiscience handheld fluorometer by Beagle Bioproducts Inc (U.S.A.) is an example of a low-cost handheld instrument for single measurements *in situ* or in the lab. It can be calibrated for the detection of phycocyanine (600 nm excitation, 650 nm emission) and/or chlorophyll (440 nm excitation, 670 nm emission). The required sample volume is 200 µL with no pre-treatment necessary. The device has USB connectivity and a LCD touch-screen display user interface. The detection range is 10 to 100 000 ppb for phycocyanine and 0.25 to 2 500 ppb for chlorophyll.

Spectral fluorometers use either fluorescence emission or fluorescence excitation. The first method provides a spectrum of algal pigment fluorescence at a single excitation wavelength. The second method varies the fluorescence excitation wavelength and provides data on the absorption of different pigments. Wet Labs (U.S.A.) has developed the ECO range of fluorescence emission fluorometers. They operate at full ocean depth and provide with data on chlorophyll-a, coloured dissolved organic matter, fluorescein/rhodamin and phycorethrin/phycocyanine.

Excitation relaxation fluorometers measure changes in chlorophyll fluorescence on a very short timescale. One type of technology is Fast Repetition Rate Fluorometers (FRRF), which used by the FastOcean family of systems (Chelsea Technologies, U.K.). They allow calculations of the rate of absolute photosynthetic electron transport on functional photosystem II per unit volume, both in the lab and *in situ* (Oxborough et al., 2012).

Bioluminescence is the natural ability to produce and emit usually blue-green light at ~500 nm by dinoflagellate phytoplankton species, like *Alexandrium tamarense* (Haddock et al., 2010). A submersible sensor for use in moorings and towed vehicles is the GlowTracka (Chelsea Technologies, U.K.). A photodiode measures the light (as photons) produced by algae in the sample as a function of time. Seawater is flowed through the detector by a 28 mm pipe and the entire sensor weighs 2.4 kg. Information on primary production of bioluminescent species is acquired and can be correlated to imaging data in order to identify the specific bioluminescent species.

Imaging

The Woods Hole Oceanographic Institute (WHOI) first developed the Video Plankton Recorder (VPR) (Davis et al., 1992). This is an underwater video microscope and has been deployed in AUVs, moorings and towed vehicles. The most recent version can image particles ranging from 100 µm to 1 cm at a field of view from 5 to 20 mm (Davis et al., 2005). The instrument is not commercially available but can be used via the Institute's pool of instrumentation.

The FlowCAM system was later developed at the Bigelow Laboratory for Ocean Sciences (U.S.A.) (Sieracki et al., 1998). The sample is drawn into a flow cell where digital images of particles are continuously acquired by a flash illumination from a Light Emitting Diode (LED). A Charge-Coupled Device (CCD) camera captures the image and image analysis software resolves the information. Fluorescence information is also recorded by a Photomultiplier Tube (PMT), although the instrument does not operate in the same principle as a flow cytometer, not having sheath flow. FlowCAM is available commercially with its most recent VS series model specification offering imaging capabilities of particles in the size range of 2 µm to 2 mm at a flow rate of 3 mL/min. The system is portable with an integrated user interface comprising of a screen, keyboard and mouse, all housed in a hard case.

Flow Cytometry and Impedance Spectroscopy

Flow cytometry is used *in situ* by transferring bench-top instruments on board research vessels or by deployable versions of the bench-top instruments, usually tethered to a ship or to a stationary platform. A flow cytometer is defined as the instrument that measures and analyses signals from particles in a liquid stream passing through a beam of light (Givan, 2001). A sheath fluid is used to align and focus the particles so that they to flow in a single line. At any one time only one particle crosses the light beam, usually from a laser source; a detector then acquires an optical or electrical impedance signal. The latter is based on the principle of impedance spectroscopy, which measures the electrical properties of particles passing through a set of electrodes while applying an AC potential.

The first cytometric instrument developed in the Netherlands for phytoplankton study was the Optical Plankton Analyzer (OPA) (Dubelaar et al., 1989; Peeters et al., 1989). It detects a larger particle size range (>1 mm) than commercial flow cytometers (1-25 µm). Other design features specific to marine applications include a large dynamic range of 1000x factor and modular excitation/detection optics. A large flow cell and high flow rates are used to achieve these specific requirements. Cytobuoy is a commercially available floating standalone version of the original OPA system (Dubelaar and Gerritzen, 2000; Dubelaar et al., 1999) and can be deployed in a buoy. A portable bench-top CytoSense and submersible CytoSub have also been developed within the same family of instruments. They are very comparable to FlowCAM because they do not require an external sheath flow like typical flow cytometers. A wide and slow-shear flow cuvette is used to focus the sample. Seawater, which has been filtered through a capsule filter, is used as sheath flow liquid. Forward scatter, side scatter, orange and red fluorescence signals are detected by photodiodes and solid-state photomultiplier tubes. Excitation is by a single low power neodymium-doped Yttrium Aluminium Garnet (Nd:YAG) laser.

At the same time, the FlowCytoBot was developed at WHOI (Olson et al., 2003). It is capable of light scattering and fluorescence particle analysis with an adjustable flow rate of 12.3 to 50 µL/min. It recycles sheath flow similarly to CytoBuoy and also uses a diode-pumped laser. An imaging FlowCytoBot was presented in 2007 (Olson and Sosik, 2007) and used to provide early warning for a *Dinophysis* bloom in the Gulf of Mexico in 2008 (Campbell et al., 2010). FlowCytoBot is optimised for smaller phytoplankton cells (1-15 µm) while its imaging version can visualise cells with larger diameters (10-300 µm). Both instruments have automated features for enumeration, identification and cell sizing and can be deployed unattended for up to six months. They include antifouling control and real-time fluidity sensing.

The main difference between the European OPA and American FlowCytoBot systems is their tethering capability. Cytobuoy uses batteries as a power supply and radio communications, whereas FlowCytoBot is a cabled instrument. Neither has been deployed on AUVs up to now. A commercial version of the imaging FlowCytoBot is now licensed from McLane Research Labs (U.S.A.).

Microfluidic flow cytometers could be deployed on AUVs, as their small size would fit the payload of the vehicles. Two lab-based micro-cytometers for phytoplankton have been published but neither has been tested on an autonomous vehicle yet. The first system to be developed was at the Morgan lab in Southampton (U.K.) for combined impedance spectroscopy and intrinsic fluorescence measurements (Benazzi et al., 2007). A PDMS chip was fabricated with a microchannel of 11-µm height and 20-µm width. Fluorescence was measured at three wavelengths with *Isochrysis galbana*, *Rhodosorus marinus* and

Synecococcus tested as exemplar species. Results from this device with non-integrated optics compared well with those acquired from a bench-top flow cytometer. However, only one-dimensional flow focusing was used (in the width direction), and impedance measurements could not size the smaller species *Synecococcus*.

At the National Research Lab, the Ligler group developed a micro-cytometer for fluorescence only measurements with two-dimensional flow focusing (Hashemi et al., 2011a; Hashemi et al., 2011b). This was achieved using chevron-shaped grooves fabricated on the micro-channel walls. The chip has integrated optics for light scattering and up to three colours of fluorescence. The channel dimensions were at 390 µm by 130 µm and the chip was also fabricated from PDMS using standard micro-fabrication procedures. Phytoplankton species *Karenia brevis*, *Synechococcus* sp., *Pseudo-Nitzchia*, and *Alexandrium*, ranging from 1 to 50 µm in diameter, were analysed. The smaller in size, *Synechococcus*, was distinguished, which was not feasible with impedance measurements by the Southampton device. However, there was some overlap between the other populations of phytoplankton, making classification difficult.

Both microsystems have potential for deployment on AUVs, if the large power requirements of the optics are improved. Another consideration is that the small size of the fluidic channels would be prone to clogging by debris and biofouling. The latter was addressed by the Southampton group, which tested different types of antifouling methods for engineering materials deployed in the marine environment (Meier et al., 2013). PDMS was a particularly bad candidate for deployable sensors, as it tended to swell in wet conditions. Chips from thermoplastics and glass had a better performance being more rigid materials.

Sensors for Speciation Using Molecular Tools

Sensors using molecular tools speciate and quantify phytoplankton by detecting nucleic acids, proteins, carbohydrates and lipids, that are characteristic for each species. Instruments in the field use: (i) direct binding of immunological (antibodies) or nucleic acid (rDNA) probes; and (ii) NAAT. Both approaches are limited to species that have already been isolated, cultured and partly or wholly sequenced. Probes can be electro-active, fluorescent or luminescent.

Three large deployable *in situ* systems have been developed for molecular detection of phytoplankton: the Environmental Sample Processor (ESP) at the Monterey Bay Aquarium Research Institute (MBARI, U.S.A.), the IISA-Gene in Japan and the Automated Microbial Genosensor (AMG) at the University of South Florida (U.S.A.).

The ESP is an autonomous and submersible system that collects, analyses and archives individual water samples in "pucks", custom-designed chambers with pre-stored reagents (Scholin et al., 1998). A rotating carousel automates the process. It weighs ~27 kg and is a cylinder with a 56-cm diameter and 82-cm height. All the steps of sample processing occur under atmospheric pressure. Toxic diatom *Pseudo-nitzschia australis* was first detected on the ESP with an automated competitive ELISA immunoassay (Doucette et al., 2009). This species produces domoic acid, a tightly monitored amnesic shellfish poison. Samples were archived and stored on filters for fluorescence *in situ* hybridization

(FISH) assays performed back in the laboratory (Greenfield et al., 2006). Fluorescent rRNA targeted DNA probes were used. The limit of detection achieved was 100 cells per litre. An additional module, the MicroFludics Block (MFB) was later developed for *in situ* solid-phase nucleic acid extraction and quantitative PCR (Preston et al., 2011). The DNA was extracted and purified from 100-1 000 mL samples down to 50 µL eluent aliquots. Real-time PCR was performed from these samples and the results were broadcasted back to shore after amplification. This version of the system was deployed for 28 days in the Monterey Bay on a coastal mooring. Meta-transcriptomic analyses were also feasible from samples collected autonomously by the ESP and preserved in ambient temperatures until retrieval (Ottesen et al., 2011). RNA samples were filtered *in situ* and treated with preservative RNAlater® (Life Technologies). The ESP is now commercially available in the \$250K price range by Spyglass Biosecurity Inc. (U.S.A.). Until now it has been deployed on AUVs, moorings, remotely operated vehicles and profiling drifters. Deep-sea deployments of the ESP were also performed using both the qPCR and rRNA assays at 4 000 m depth in the Santa Monica Basin (Ussler et al., 2013).

The IISA-Gene system uses microfluidic components that operate at ambient pressure for *in situ* PCR measurements (Fukuba et al., 2011). The sensor is very versatile because of the miniaturised components, and can operate at extreme depths. The only drawback is that it collects relatively small sample volumes (0.5 mL per hour). All components of the analysis unit are immersed in fluorinated oil and are connected by underwater cables to a remote operator on a surface vessel. The most recent version of the instrument has 128 sampling channels for in-line parallel collection (Okamura et al., 2014).

The AMG is a deployable system which uses an isothermal NAAT, namely Nucleic Acid Sequence-Based Amplification (NASBA) (Fries et al., 2007). It is designed for long-term *in situ* monitoring of *Karenia brevis* using sample collection/filtration, RNA purification/ concentration and real-time RNA amplification. *K. brevis* produces potent lipophilic bio-toxins which are a threat to local fauna and humans. The analyser measures 22 cm in diameter and 122 cm in length, and has integrated custom-made Printed Circuit Boards (PCBs), reagent bags, fluidic actuators (valves, pumps and manifolds), as well as a battery pack. Deployment is possible for up to three days with collection of 30-50 mL seawater samples and completion of 12 NASBA assays. Results can be transmitted back to the shore in real-time. A handheld version of AMG has been developed by the same lab for field detection of *K. brevis* (Casper et al., 2007). It features integrated RNA extraction and purification by silica columns and real-time NASBA using molecular beacons.

A smaller in size microfluidic real-time NASBA microsystem was developed at the University of Southampton (U.K.) as part of European-Commission-funded project LABONFOIL. The first version of the sensor with non-integrated optics used acrylic microchips fabricated by rapid prototyping (Tsaloglou et al., 2011). RNA from 10 *K. brevis* cells was detected within 3 minutes. A second version used disposable injection-moulded cyclic olefin co-polymer chips and a portable reader (20 cm height, 30 cm width, 20 cm depth). All the NASBA reagents were preserved on-chip for up to eight months in room temperature. Additional functionality like cell lysis on-chip has been demonstrated separately: electroporation of *K. brevis* cysts (Bahi et al., 2011) and ultra-sonication of

Pseudo-nitzschia (Wu et al., 2011). None of these systems have been deployed in the field yet. However, they all contribute with features such as quick time-to-result, reagent preservation and a small payload.

Analysers for biological measurements have also been developed for detection of pathogenic organisms in water. Although not directly targeting algae, they have comparable components, which could be adapted for phytoplankton microbial ecology studies. A few examples will be highlighted here: two small portable and a large deployable system.

The Delattre group in France developed a portable integrated system for the detection of *E.coli* in water (Delattre et al., 2012). It measures 30 cm by 50 cm by 40 cm and requires a sample volume of up to 10 mL. Cells are captured on anion-exchange magnetic beads, resulting to a 250x concentration factor. Cells are lysed on the device and their DNA is extracted for subsequent amplification. Conventional pumps and valves are used for fluidic control. The pure DNA eluent is transferred to an EWOD microfluidic chip where multiplex PCR takes places in 64 nL volume reactions in discrete droplets.

The United States Naval Research Lab (NRL) developed the Array Biosensor (Ligler et al., 2007). It is a portable immunosensor for the detection of six different analytes using competitive in-line immunoassays. The reactions are automated and fluidic control is achieved by commercial peristaltic valves and solenoid valves. The reagents are stored on the device which weighs ~7kg. Chips are comprised of glass and PDMS and fluorescence is detected using an integrated CCD camera. The Array Biosensor is now licensed to several companies, which produce it commercially for different applications.

The Biohazard Water Analyser by Early Warning Inc., a spin-off from NASA's Ames Research Center, is a large deployable system for detection of *E.coli*, Legionella, Salmonella and up to 25 pathogenic bacteria, viruses and parasites. Cells are concentrated on the device, multiplex DNA is performed on an electrochemical biosensor and the results are transmitted off-site. There is a separate working electrode for each pathogen with species-specific DNA probes (~30 bp) immobilised on nanowires at the surface of the electrode. The probes are complementary to rRNA sequence for each target. Upon hybridisation of the rRNA target, current is measurable on the electrodes upon application of a 1.05 V potential across the electrodes, caused by the reaction of guanine with cytosine. The magnitude of the electrical current measured is proportional to the amount of pathogenic rRNA in the sample.

Conclusions

Monitoring of phytoplankton in the oceans provides with insight into biogeochemical cycles, microbial ecology, pollution and climate change. Data collection relies mostly on data collected during research cruises, Eulerian platforms and AUVs. Existing *in situ* systems are very large, expensive, rely on remote control and can only take semi-continuous measurements. Advances in microfluidics can provide solutions for smaller deployable *in situ* platforms. These have potential for long-term deployment as networks of mobile Lagrangian sensors. Such networks would improve spatial resolution of existing data collection and combined with deployed biological microsensors on AUVs for real-time data could contribute to our knowledge and understanding of changes in phytoplankton populations in the oceans.

Future trends

The European Commission made a large investment as part of the "The Ocean of Tomorrow" initiative. A total of €195.6 million in 2010-2013 were awarded to 31 separate projects with academic and industrial partners. The projects span from developing offshore platforms, fundamental studies of marine microbial diversity, antifouling strategies to *in situ* marine sensing technologies (including biosensors). At least three of the projects, SCHeMA, Sea-on-a-Chip and SenseOCEAN, are developing microfluidic sensors with molecular sensing of phytoplankton.

Advances in the fields of biotechnology and miniaturised optics, electronics and fluidics will continue to drive the field of miniaturised sensors, which will surely benefit the development of sensors for environmental applications.

Web resources

- Biomark HD System (Fluidigm) http://vimeo.com/14655710
- Phylochip http://www2.lbl.gov/tt/techs/lbnl2229.html
- GeoChip http://www.glomics.com/gch-tech.html#
- NASA, Earth Observatory http://earthobservatory.nasa.gov/Features/Phytoplankton/
- Smithsonian Environmental Centre http://www.serc.si.edu/labs/phytoplankton/guide/
- FlowCAM commercial site http://www.fluidimaging.com/products-portable.htm
- Video Plankton Recorder http://www.whoi.edu/instruments/viewInstrument.do?id=1007
- CytoBuoy http://www.cytobuoy.com/products/submersible/
- Imaging FlowCytobot commercial site http://www.mclanelabs.com/master_page/product-type/samplers/imaging-flowcytobot
- Beagle Amiscience fluorometer https://beaglebioproducts.com/wp-content/uploads/2014/04/Fluorometer-Application-Note.pdf
- ECO FL spectral fluorometer http://www.wetlabs.com/eco-fl
- FastOcean FRR Fluorometer http://www.chelsea.co.uk/marine/fluorometers/fastocean-systems
- GlowTracka bioluminescence sensor http://www.chelsea.co.uk/allproduct/environmentalfresh-water/sensors/glowtracka-bioluminescence
- ESP http://www.spyglassbio.com
- Biohazard Water Analyser http://www.earlywarninginc.com/

Ackowlegdgements

The author would like to acknowledge financial support from the European Commission Marie Curie Actions as an International Outgoing Marie Curie Fellowship (Grant No. PIOF-GA-2013-622879-ENVIRO).

References

Adornato, L., Cardenas-Valencia, A., Kaltenbacher, E., Byrne, R.H., Daly, K., Larkin, K., Hartman, S., Mowlem, M., Prien, R.D., and Garcon, V. (2010). *In Situ* nutrient sensors for ocean observing systems. In Proceedings of OceanObs '09: Sustained Ocean Observations and Information for Society, J. Hall, D.E. Harrison, and D. Stammer, eds. (Venice, Italy: ESA Publication WPP-306), pp. 1-10.

Airs, R.L., and Garrido, J.L. (2011). Liquid chromatography-mass spectrometry for pigment analysis. In Phytoplankton Pigments: Characterization, Chemotaxonomy and Applications in Oceanography, (Cambridge, U.K.: Cambridge University Press), pp. 314-342.

Bahi, M.M., Tsaloglou, M.-N., Mowlem, M., and Morgan, H. (2011). Electroporation and lysis of marine microalga *Karenia brevis* for RNA extraction and amplification. J. Roy. Soc. Interface *8*, 601-608.

Benazzi, G., Holmes, D., Sun, T., Mowlem, M.C., and Morgan, H. (2007). Discrimination and analysis of phytoplankton using a microfluidic cytometer. IET Nanobiotechnology *1*, 94-101.

Bidigare, R.R., Van Heukelem, L., and Trees, C.C. (2005). Analysis of algal pigments by high-performance liquid chromatohraphy. In Algal Culturing Techniques, R.A. Andersen, ed. (New York, U. S. A. : Elsevier Academic Press), pp. 327-346.

Blondeau-Patissier, D., Gower, J.F.R., Dekker, A.G., Phinn, S.R., and Brando, V.E. (2014). A review of ocean color remote sensing methods and statistical techniques for the detection, mapping and analysis of phytoplankton blooms in coastal and open oceans. Progr. Oceanogr. *123*, 123-144.

Campbell, L., Olson, R.J., Sosik, H.M., Abraham, A., Henrichs, D.W., Hyatt, C.J., and Buskey, E.J. (2010). First harmful *Dinophysis* (Dinophyceae, Dinophysiales) bloom in the U.S. is revealed by automated imaging flow cytometry J. Phycol. *46*, 66-75.

Casper, E.T., Patterson, S.S., Bhanushali, P., Farmer, A., Smith, M., Fries, D.P., and Paul, J.H. (2007). A handheld NASBA analyzer for the field detection and quantification of Karenia brevis. Harmful Algae *6*, 112-118.

Choi, K., Ng, A.H.C., Fobel, R., and Wheeler, A.R. (2012). Digital microfluidics. Annu. Rev. Anal. Chem. *5*, 413-440.

Crevillén, A.G., Hervás, M., López, M.A., González, M.C., and Escarpa, A. (2007). Real sample analysis on microfluidic devices. Talanta *74*, 342-357.

Davis, C.S., Gallager, S.M., Berman, M.S., Haury, L.R., and Strickler, J.R. (1992). The video plankton recorder (VPR): design and initial results. Arch. Hydrobiol. Beih. *36*, 67-81.

Davis, C.S., Thwaites, F.T., Gallager, S.M., and Hu, Q. (2005). A three-axis fast-tow digital Video Plankton Recorder for rapid surveys of plankton taxa and hydrography. Limnol. Oceanogr.-Meth. *3*, 59-74.

Davis, K.M., Miura, K., Sugimoto, N., and Hirao, K. (1996). Writing waveguides in glass with a femtosecond laser. Opt. Lett. *21*, 1729-1731.

Delattre, C., Allier, C.P., Fouillet, Y., Jary, D., Bottausci, F., Bouvier, D., Delapierre, G., Quinaud, M., Rival, A., Davoust, L., *et al.* (2012). Macro to microfluidics system for biological environmental monitoring. Biosens. Bioelectron. *36*, 230-235.

Delauney, L., Compere, C., and Lehaitre, M. (2010). Biofouling protection for marine environmental sensors. Ocean Sci. *6*, 503-511.

DeSantis, T., Brodie, E., Moberg, J., Zubieta, I., Piceno, Y., and Andersen, G. (2007). High-density universal 16S rRNA microarray analysis reveals broader diversity than typical clone library when sampling the environment. Microb. Ecol. *53*, 371-383.

Doucette, G.J., Mikulski, C.M., Jones, K.L., King, K.L., Greenfield, D.I., Marin, R., Jensen, S., Roman, B., Elliott, C.T., and Scholin, C.A. (2009). Remote, subsurface detection of the algal toxin domoic acid onboard the Environmental Sample Processor: Assay development and field trials. Harmful Algae *8*, 880-888.

Dubelaar, G.B.J., and Gerritzen, P.L. (2000). CytoBuoy: a step forward towards using flow cytometry in operational oceanography. Sci. Mar. *64*, 255-265.

Dubelaar, G.B.J., Gerritzen, P.L., Beeker, A.E.R., Jonker, R.R., and Tangen, K. (1999). Design and first results of CytoBuoy: A wireless flow cytometer for *in situ* analysis of marine and fresh waters. Cytometry *37*, 247-254.

Dubelaar, G.B.J., Groenewegen, A.C., Stokdijk, W., Van Den Engh, G.J., and Visser, J.W.M. (1989). Optical plankton analyser: A flow cytometer for plankton analysis, II: Specifications. Cytometry *10*, 529-539.

Embleton, K.V., Gibson, C.E., and Heaney, S.I. (2003). Automated counting of phytoplankton by pattern recognition: a comparison with a manual counting method. J. Plankton Res. *25*, 669-681.

Erickson, J.S., Hashemi, N., Sullivan, J.M., Weidemann, A.D., and Ligler, F.S. (2011). *In situ* phytoplankton analysis: There's plenty of room at the bottom. Anal. Chem. *84*, 839-850.

Fair, R.B., Khlystov, A., Tailor, T.D., Ivanov, V., Evans, R.D., Griffin, P.B., Srinivasan, V., Pamula, V.K., Pollack, M.G., and Zhou, J. (2007). Chemical and biological applications of digital-microfluidic devices. IEEE Des. Test Comput. *24*, 10-24.

Fries, D., Paul, J., Smith, M., Farmer, A., Casper, E., and Wilson, J. (2007). The Autonomous Microbial Genosensor, an *in situ* sensor for marine microbe detection. Microsc. Microanal. *13*, 514-515.

Fukuba, T., Miyaji, A., Okamoto, T., Yamamoto, T., Kaneda, S., and Fujii, T. (2011). Integrated *in situ* genetic analyzer for microbiology in extreme environments. RSC Advances *1*, 1567-1573.

Givan, A.L. (2001). Instrumentation: into the black box. In Flow Cytometry: First Principles (Oxford, U.K.: John Wiley and Sons, Inc.), pp. 15-39.

Greenfield, D.I., Marin, R., Jensen, S., Massion, E., Roman, B., Feldman, J., and Scholin, C.A. (2006). Application of environmental sample processor (ESP) methodology for quantifying *Pseudo-nitzschia australis* using ribosomal RNA-targeted probes in sandwich and fluorescent *in situ* hybridization formats. Limnol. Oceanogr.-Meth. *4*, 426-435.

Haddock, S.H.D., Moline, M.A., and Case, J.F. (2010). Bioluminescence in the sea. Annu. Rev. Mar. Sci. *2*, 443-493.

Hansen, C.L., Skordalakes, E., Berger, J.M., and Quake, S.R. (2002). A robust and scalable microfluidic metering method that allows protein crystal growth by free interface diffusion. Proc. Natl. Acad. Sci. U.S.A. *99*, 16531-16536.

Hashemi, N., Erickson, J.S., Golden, J.P., Jackson, K.M., and Ligler, F.S. (2011a). Microflow cytometer for optical analysis of phytoplankton. Biosens. and Bioelectr. *26*, 4263-4269.

Hashemi, N., Erickson, J.S., Golden, J.P., and Ligler, F.S. (2011b). Optofluidic characterization of marine algae using a microflow cytometer. Biomicrofluidics *5*, 032009.

He, Z., Gentry, T.J., Schadt, C.W., Wu, L., Liebich, J., Chong, S.C., Huang, Z., Wu, W., Gu, B., Jardine, P., *et al.* (2007). GeoChip: a comprehensive microarray for investigating biogeochemical, ecological and environmental processes. ISME J. *1*, 67-77.

Heidelberg, K.B., and Heidelberg, J.F. (2005). Marine environmental genomics: New secrets from a mysterious ocean. Mar. Technol. Soc. J. *39*, 94-98.

Horsman, K.M., Bienvenue, J.M., Blasier, K.R., and Landers, J.P. (2007). Forensic DNA analysis on microfluidic devices: A review. J. Forensic Sci. *52*, 784-799.

Huang, Y., Joo, S., Duhon, M., Heller, M., Wallace, B., and Xu, X. (2002). Dielectrophoretic cell separation and gene expression profiling on microelectronic chip arrays. Anal. Chem. *74*, 3362-3371.

Huang, Y.Y., Beal, C.M., Cai, W.W., Ruoff, R.S., and Terentjev, E.M. (2010). Micro-Raman spectroscopy of algae: Composition analysis and fluorescence background behavior. Biotechn. Bioeng. *105*, 889-898.

Irigoien, X., Meyer, B., Harris, R., and Harbour, D. (2004). Using HPLC pigment analysis to investigate phytoplankton taxonomy: the importance of knowing your species. Helgol. Mar. Res. *58*, 77-82.

Jeffrey, S.W., Wright, S., and Zapata, M. (2011). Microalgal classes and their signature pigments. In Phytoplankton Pigments: Characterization, Chemotaxonomy and Applications in Oceanography, S. Roy, C.A. Jewelly, E.S. Egeland, and G. Johnsen, eds. (Cambridge, U.K.: Cambridge University Press), pp. 3-77.

Jones, B., Edwards, R., Skipp, P., O'Connor, C.D., and Iglesias-Rodriguez, M.D. (2011). Shotgun proteomic analysis of *Emiliania huxleyi*, a marine phytoplankton species of major biogeochemical importance. Mar. Biotechnol. *13*, 496-504.

Khan, I.U., Serra, C.A., Anton, N., and Vandamme, T. (2013). Microfluidics: A focus on improved cancer targeted drug delivery systems. J. Control. Rel. *172*, 1065-1074.

Kim, S., Huang, B., and Zare, R.N. (2007). Microfluidic separation and capture of analytes for single-molecule spectroscopy. Lab on a Chip *7*, 1663-1665.

Kirkpatrick, G.J., Millie, D.F., Moline, M.A., Lohrenz, S.E., and Schofield, O.M. (2011). Automated, in-water determination of colored dissolved organic material and phytoplankton community structure using the optical phytoplankton discriminator. Ocean Sensing and Monitoring III *8030*, 9.

Labuz, J.M., and Takayama, S. (2014). Elevating sampling. Lab on a Chip *14*, 3165-3171.

Lampitt, R.S., Favali, P., Barnes, C.R., Church, M.J., Cronin, M.F., Hill, K.L., Kaneda, Y., Karl, D.M., Knap, A.H., McPhaden, M.J., *et al.* (2010). *In situ* sustained Eulerian observatories. In Proceedings of OceanObs '09: Sustained Ocean Observations and Information for Society, J. Hall, D.E. Harrison, and D. Stammer, eds. (Venice, Italy: ESA Publication WPP-306), pp. 27-37.

Li, H.-F., and Lin, J.-M. (2009). Applications of micro-fluidic systems in environmental analysis. Anal. Bioanal. Chem. *393*, 555-567.

Li, S., Xu, J., Chen, J., Chen, J., Zhou, C., and Yan, X. (2014). The major lipid changes of some important diet microalgae during the entire growth phase. Aquaculture *428–429*, 104-110.

Ligler, F.S., Sapsford, K.E., Golden, J.P., Shriver-Lake, L.C., Taitt, C.R., Dyer, M.A., Barone, S., and Myatt, C.J. (2007). The array biosensor: Portable, automated systems. Anal. Sci. *23*, 5-10.

Manz, A., Graber, N., and Widmer, H.M. (1990). Miniaturized total chemical analysis systems: A novel concept for chemical sensing. Sensor. Actuat. B- Chem. *1*, 244-248.

Mariella, R. (2008). Sample preparation: the weak link in microfluidics-based biodetection. Biomed. Microdevices *10*, 777-784.

Mark, D., Haeberle, S., Roth, G., von Stetten, F., and Zengerle, R. (2010). Microfluidic lab-on-a-chip platforms: requirements, characteristics and applications. Chem. Soc. Rev. *39*, 1153-1182.

Meier, A., Tsaloglou, N.M., Mowlem, M.C., Keevil, C.W., and Connelly, D.P. (2013). Hyperbaric biofilms on engineering surfaces formed in the deep sea. Biofouling *29*, 1029-1042.

Mills, G., and Fones, G. (2012). A review of *in situ* methods and sensors for monitoring the marine environment. Sens. Rev. *32*, 17-28.

Moore, C., Barnard, A., Fietzek, P., Lewis, M.R., Sosik, H.M., White, S., and Zielinski, O. (2009). Optical tools for ocean monitoring and research. Ocean Sci. *5*, 661-684.

Morgan, H., Hughes, M.P., and Green, N.G. (1999). Separation of submicron bioparticles by dielectrophoresis. Biophys. J *77*, 516-525.

Mullis, K.B., Faloona, F.A., and Ray, W. (1987). Specific synthesis of DNA in vitro via a polymerase-catalyzed chain reaction. In Methods in Enzymology (New York, U. S. A.: Academic Press), pp. 335-350.

Nge, P.N., Rogers, C.I., and Woolley, A.T. (2013). Advances in microfluidic materials, functions, integration, and applications. Chem. Rev. *113*, 2550-2583.

Okamura, K., Noguchi, T., Hatta, M., Sunamura, M., Suzue, T., Kimoto, H., Fukuba, T., and Fujii, T. (2014). Development of a 128-channel multi-water-sampling system for underwater platforms and its application to chemical and biological monitoring. Meth. Oceanogr. *8*, 75-90.

Olson, R.J., Shalapyonok, A., and Sosik, H.M. (2003). An automated submersible flow cytometer for analyzing pico- and nanophytoplankton: FlowCytobot. Deep-Sea Res. Pt I *50*, 301-315.

Olson, R.J., and Sosik, H.M. (2007). A submersible imaging-in-flow instrument to analyze nano- and microplankton: Imaging FlowCytobot. Oceanogr.and Limnol. Meth. *5*, 195-203.

Olson, R.J., Vaulot, D., and Chisholm, S.W. (1985). Marine phytoplankton distributions measured using shipboard flow cytometry. Deep-Sea Res. Pt I *32*, 1273-1280.

Ottesen, E.A., Marin, R., III, Preston, C.M., Young, C.R., Ryan, J.P., Scholin, C.A., and DeLong, E.F. (2011). Metatranscriptomic analysis of autonomously collected and preserved marine bacterioplankton. I.S.M.E. J. *5*, 1881-1895.

Oxborough, K., Moore, C.M., Suggett, D.J., Lawson, T., Chan, H.G., and Geider, R.J. (2012). Direct estimation of functional PSII reaction center concentration and PSII electron flux on a volume basis: a new approach to the analysis of Fast Repetition Rate Fluorometry (FRRF) data. Limnol. Oceanogr.-Meth. *10*, 142-154.

Peeters, J.C.H., Dubelaar, G.B.J., Ringelberg, J., and Visser, J.W.M. (1989). Optical plankton analyser: A flow cytometer for plankton analysis, I: Design considerations. Cytometry *10*, 522-528.

Pilát, Z., Bernatová, S., Ježek, J., Šerý, M., Samek, O., Zemánek, P., Nedbal, L., and Trtílek, M. (2012). Raman microspectroscopy of algal lipid bodies: β-carotene quantification. J Appl Phycol *24*, 541-546.

Preston, C.M., Harris, A., Ryan, J.P., Roman, B., Marin, R., Jensen, S., Everlove, C., Birch, J., Dzenitis, J.M., Pargett, D., *et al.* (2011). Underwater application of quantitative PCR on an ocean mooring. Plos One *6*, e22522.

Robbins, I.C., Kirkpatrick, G.J., Blackwell, S.M., Hillier, J., Knight, C.A., and Moline, M.A. (2006). Improved monitoring of HABs using autonomous underwater vehicles (AUV). Harmful Algae *5*, 749-761.

Roussel, B. (2013). Microfluidic Applications Market 2013 Report (Yole Developpement).

Rusch, D.B., Halpern, A.L., Sutton, G., Heidelberg, K.B., Williamson, S., Yooseph, S., Wu, D.Y., Eisen, J.A., Hoffman, J.M., Remington, K., *et al.* (2007). The Sorcerer II global ocean sampling expedition: Northwest Atlantic through Eastern Tropical Pacific. PLoS Biol *5*, 398-431.

Rynearson, T.A., and Palenik, B. (2011). Learning to read the oceans: Genomics of marine phytoplankton. In Advances in Marine Biology, L. Michael, ed. (New York, U. S. A.: Academic Press), pp. 1-39.

Schaap, A., Rohrlack, T., and Bellouard, Y. (2012a). Lab on a chip technologies for algae detection: a review. J. Biophotonics *5*, 661-672.

Schaap, A., Rohrlack, T., and Bellouard, Y. (2012b). Optical classification of algae species with a glass lab-on-a-chip. Lab on a Chip *12*, 1527-1532.

Schofield, O., Glenn, S., and Moline, M. (2013). The Robot ocean network. American Scientist *101*, 434-441.

Scholin, C., Massion, G., Mellinger, E., Brown, M., Wright, D., and Cline, D. (1998). The development and application of molecular probes and novel instrumentation for detection of harmful algae. Paper presented at: Ocean Community Conference '98 (Marine Technology Society).

Schulze, K., Tillich, U.M., Dandekar, T., and Frohme, M. (2013). PlanktoVision - an automated analysis system for the identification of phytoplankton. BMC Bioinformatics *14*, 10.

Shade, A., Carey, C.C., Kara, E., Bertilsson, S., McMahon, K.D., and Smith, M.C. (2009). Can the black box be cracked? The augmentation of microbial ecology by high-resolution, automated sensing technologies. ISME J. *3*, 881-888.

Sieracki, C.K., Sieracki, M.E., and Yentsch, C.S. (1998). An imaging-in-flow system for automated analysis of marine microplankton. Mar. Ecol. Progr. Series *168*, 285-296.

Siggins, A., Gunnigle, E., and Abram, F. (2012). Exploring mixed microbial community functioning: recent advances in metaproteomics. FEMS Microbiol. Ecol. *80*, 265-280.

Sosik, H., Olson, R., and Armbrust, E.V. (2010). Flow cytometry in phytoplankton research. In Chlorophyll a Fluorescence in Aquatic Sciences: Methods and Applications, D.J. Suggett, O. Prášil, and M.A. Borowitzka, eds. (Rotterdam, Netherlands: Springer), pp. 171-185.

Tang, S.K.Y., Stan, C.A., and Whitesides, G.M. (2008). Dynamically reconfigurable liquid-core liquid-cladding lens in a microfluidic channel. Lab on a Chip *8*, 395-401.

Teh, S.Y., Lin, R., Hung, L.H., and Lee, A.P. (2008). Droplet microfluidics. Lab on a Chip *8*, 198-220.

Tomas, C.R. (1997). Identifying marine phytoplankton (San Diego, California, U.S.A.: Academic Press, Inc.).

Tsaloglou, M.-N., Bahi, M.M., Waugh, E.M., Morgan, H., and Mowlem, M. (2011). On-chip real-time nucleic acid sequence-based amplification for RNA detection and amplification. Anal. Methods *3*, 2127-2133.

Ussler, W., Preston, C., Tavormina, P., Pargett, D., Jensen, S., Roman, B., Marin, R., Shah, S.R., Girguis, P.R., Birch, J.M., *et al.* (2013). Autonomous application of quantitative PCR in the deep sea: *In situ* surveys of aerobic methanotrophs using the deep-sea Environmental Sample Processor. Env. Sci. and Techn. *47*, 9339-9346.

van Doorn, R., Klerks, M.M., van Gent-Pelzer, M.P.E., Speksnijder, A.G.C.L., Kowalchuk, G.A., and Schoen, C.D. (2009). Accurate quantification of microorganisms in PCR-inhibiting environmental DNA extracts by a novel internal amplification control approach using biotrove open arrays. Appl. Env. Microbiol. *75*, 7253-7260.

Whitesides, G. (2014). The lab finally comes to the chip! Lab on a Chip *14*, 3125-3126.

Whitesides, G.M. (2006). The origins and the future of microfluidics. Nature *442*, 368-373.

Williamson, S.J., Allen, L.Z., Lorenzi, H.A., Fadrosh, D.W., Brami, D., Thiagarajan, M., McCrow, J.P., Tovchigrechko, A., Yooseph, S., and Venter, J.C. (2012). Metagenomic exploration of viruses throughout the Indian Ocean. PLoS ONE *7*, e42047.

Wu, C., Lillehoj, P.B., Sabet, L., Wang, P., and Ho, C.-M. (2011). Ultrasonication on a microfluidic chip to lyse single and multiple Pseudo-nitzschia for marine biotoxin analysis. Biotechn. J. *6*, 150-155.

Wu, J., and Gu, M. (2011). Microfluidic sensing: state of the art fabrication and detection techniques. J. Biomed. Opt. *16*, 12.

Yan, L., Zhou, J., Zheng, Y., Gamson, A.S., Roembke, B.T., Nakayama, S., and Sintim, H.O. (2014). Isothermal amplified detection of DNA and RNA. Mol. Biosystems *10*, 970-1003.

Zehr, J.P., Hewson, I., and Moisander, P. (2009). Molecular biology techniques and applications for ocean sensing. Ocean Sci. *5*, 101-113.

Zeng, Y., and Wang, T. (2013). Quantitative microfluidic biomolecular analysis for systems biology and medicine. Anal. Bioanal. Chem. *405*, 5743-5758.

Other Publications of Interest

Flores, E. and Herrero, A. (2014). The Cell Biology of Cyanobacteria. Caister Academic Press, U.K. ISBN: 978-1-908230-38-6

Herold, K.E. and Rasooly, A. (2009). Lab-on-a-Chip Technology: Fabrication and Microfluidics. Caister Academic Press, U.K. ISBN: 978-1-904455-46-2

Herold, K.E. and Rasooly, A. (2009) Lab-on-a-Chip Technology: Biomolecular Separation and Analysis. Caister Academic Press, U.K. ISBN: 978-1-904455-47-9

Wilkinson, M.G. (2015). Flow Cytometry in Microbiology: Technology and Applications. Caister Academic Press, U.K. ISBN: 978-1-910190-11-1

Xuefeng, L. (2014). Biofuels: From Microbes to Molecules. Caister Academic Press, U.K. ISBN: 978-1-908230-45-4

Other books of interest:

Climate Change and Microbial Ecology

Current Research and Future Trends
Edited by: J Marxsen
c. 220 pp, March 2016
Book: ISBN 978-1-910190-31-9 £159/$319
Ebook: ISBN 978-1-910190-32-6 £159/$319
The most important hot-topics in the area of climate change and microbial ecology.

Biofilms in Bioremediation

Current Research and Emerging Technologies
Edited by: G Lear
c. 250 pp, March 2016
Book: ISBN 978-1-910190-29-6 £159/$319
Ebook: ISBN 978-1-910190-30-2 £159/$319
Up-to-date and authoritative reviews of the latest scientific research.

Aquatic Biofilms

Ecology, Water Quality and Wastewater Treatment
Edited by: AM Romaní, H Guasch, MD Balaguer
c. 200 pp, January 2016
Book: ISBN 978-1-910190-17-3 £159/$319
Ebook: ISBN 978-1-910190-18-0 £159/$319
Characteristics and ecology of biofilms in natural ecosystems and applications linked to water pollution problems.

Flow Cytometry in Microbiology

Technology and Applications
Edited by: MG Wilkinson
xii + 218 pp, September 2015
Book: ISBN 978-1-910190-11-1 £159/$319
Ebook: ISBN 978-1-910190-12-8 £159/$319
A thorough description of flow cytometry with practical and up-to-date information aimed at microbiologists.

Biofuels

From Microbes to Molecules
Edited by: X Lu
x + 248 pp, July 2014
Book: ISBN 978-1-908230-45-4 £159/$319
Ebook: ISBN 978-1-908230-63-8 £159/$319
"a timely overview" (Biotechnol. Agron. Soc. Environ.); "a highly recommended read" (Front. Bioeng. Biotechnol.)

The Cell Biology of Cyanobacteria

Edited by: E Flores, A Herrero
x + 308 pp, May 2014
Book: ISBN 978-1-908230-38-6 £159/$319
Ebook: ISBN 978-1-908230-92-8 £159/$319
"up-to-date information" Book News; "I recommend this book" (Biospektrum)

Microbial Ecological Theory

Current Perspectives
Edited by: LA Ogilvie, PR Hirsch
viii + 112 pp, September 2012
Book: ISBN 978-1-908230-09-6 £120/$240
Ebook: ISBN 978-1-908230-71-3 £120/$240
"thought-provoking ... a useful reference book" (Microbiol. Today); "insightful contributions" (ASM Microbe)

CPSIA information can be obtained at www.ICGtesting.com
Printed in the USA
BVOW11s2103070815

412275BV00002B/2/P